BRIGHT HOLE
COSMOS
AND MULTIBANG DYNAMICS

BRIGHT HOLE
COSMOS
AND MULTIBANG DYNAMICS

ANDRE TREPANIER

Rev. Date: 06/25/2013

To order additional copies of this book, contact:
Xlibris LLC
1-888-795-4274
www.Xlibris.com
Orders@Xlibris.com
123184

CONTENTS

ACKNOWLEDGEMENTS

This book is dedicated to my mediterranean wife, Khadija, for her patience and warm presence.

Thanks to my family and all my friends from the class of '63 who encouraged me in writing 'Bright Hole Cosmos' and especially to Robert Gagnon who made many useful comments on Galactic Dynamics in Part 6.

A special thought for my friends from the Foundation set up by the late Roger Lebeuf who wrote a basic philosophical synthesis titled 'Cosmic Presence'.

Cover picture:

A transparent shell remnant from supernova SNR 0509 taken by NASA's Hubble Space Telescope is shown. A white dwarf or neutron star left over from the progenitor star is visible in the center and a shell expansion velocity of 5000 km/sec has been measured.

PREFACE

A few years ago, as I was still active as a petroleum geophysicist, I followed an introductory course on cosmology given by Hubert Reeves on the education television channel in Québec. The subject was extremely well structured and clearly presented by this expert, but I was astonished to discover the lack of scientific proofs on which the whole castle of modern cosmology is built [21]. Now retired from my field of activity, I have read and reflected for the past few years on this passionate subject.

My goal is to challenge some of the main tenets of present-day cosmology, not as an expert but simply as somebody questioning some of the shocking conclusions that have been transformed, in the course of about one century, from paradigm into pseudoscientific dogmas. The challenge will be conducted in two parts: first, current mainstream ideas will be presented with their acknowledged scientific input and with their questionable models and results; second, I will present a defendable alternative model with all due respect to physical laws and show its more universal value and consistency.

Why should we question such a complex subject where experts keep learning every day, and what could we possibly add that has not already been said? Cosmology is so fundamental that it reaches to our wonderment and deepest feelings about nature and our own cosmic presence. With new technologies and the load of digitized data from ground—and satellite-based telescopes and accessory instruments, headways have been made with factual data interpretation. But the cosmos is not a private hunting ground and does not belong to a particular club of observers. It is permanently available to anyone who wants to explore it. Like so many scientific ventures, it is based on hypothetical models that are continuously adjusted to fit new observational data. Thus, we are responsible to challenge the current big bang model and propose some improvements where we see fit, all the way from early expansion to final outcome theories.

The title *Bright Hole Cosmos* brings out the similitude with the black hole concept where a mass, within the Schwarzschild radius, is contracted by a gravitational field so intense that almost nothing can escape, not even light. It is "bright" since we are observing the universe from within and since it is filled with energy of all kinds that we can tune in. There is no outside, no observer studying our cosmos from a separate vantage point since an outsider would be unable to illuminate it or to capture any of its energy.

So I would like to share with you the following ideas and much more: the unique big bang model is replaced by a multibang model made up of many local and permanently renewed maxibangs. Universal-scale cosmic expansion is challenged by a local-scale expansion and contraction model where expansion is the dominant volumetric expression. Universal cosmological redshift is replaced by local cosmological space expansion redshift, neutral shift, and contraction blueshift, and Hubble's constant is explained as the average rate of expansion computed from a multitude of local redshifts, which add up to a larger value than the sum of the local blueshifts or neutral shifts. The cosmic model that will emerge includes time and space as permanently renewed local features of the electronuclear realm within a finite spheroidal cosmos where the conservative force of gravity, freed from the limited speed of light, becomes the great watchmaker rewinding the cosmos.

In the text, I will frequently use the scientific notation in which numbers are traditionally expressed as exponents of base 10; for example, 10^3 is 1,000, where the exponent indicates by how many digits the decimal point in 1.0 is moved to the right, and a fraction like 10^{-3} is 0.001, where the exponent indicates how many times or digits the decimal point in 1.0 is moved to the left. Note that this is different from most handheld calculators' notation, where numbers are expressed as exponents of 1, like 1^3 is 1,000, and also contrary to computer spreadsheets' notation, where the more explicit value 1E+3, representing 1,000, will be used in data tables.

Although not specifically shown, force, velocity, and acceleration are vectors with magnitude and direction while time, mass, and distance units are scalars expressed with magnitude only. References to authors or complementary articles listed in the bibliography are shown with bracketed numbers [00] in the text.

1. Introduction and Overview

Let us start this cosmic exploration by a brief historical review and a summary of the main theories and associated problems with present-day cosmological models, including the prevailing big bang paradigm and the less publicized steady state or static universe models [14]. We will apply Occam's razor principle throughout the text to facilitate the understanding of our own model and consequently present it with the least assumptions.

1.1 Introduction

Vesto Slipher may be considered as the father of galactic Doppler shift measurements, which he started in 1912 on so-called nebulae since separate galaxies were not yet identified. He later discovered that most of these nebulae were redshifted and thus receding from Earth. At that time, galaxies were known from the 100-year-old Messier catalog as nebulae, and since the cosmic ladder of distances was not yet invented, it was generally believed that these nebulae were within or in proximity to the Milky Way.

In 1922, Alexander Friedmann derived his own equations from Albert Einstein's equation of general relativity from which he showed that the universe could be expanding, although a static universe was the privileged model in those days and it would be difficult to convince the scientific community otherwise [62].

Only two years later, Edwin Hubble started measurements on Cepheid variable stars and determined that galaxies were located at great distances outside our Milky Way galaxy. In 1927, Georges Lemaître presented the idea that recession of the "nebulae" found from redshift is due to the expansion of the universe. Then, in 1929, Hubble discovered the well-known Hubble's law based on redshift that correlates distance and recession velocity although the original version was a very poor correlation with a large error in distances. Since the universe may be expanding, Lemaître suggested in 1931 to go back in time to discover the birth of the cosmos from a small point, a "primeval atom," marking the beginning of space and time.

On the other hand, Friedmann proposed an oscillatory universe model, with expansion followed by contraction. Fritz Zwicky came up with the tired light model, in which light traveling long distances loses gradually some of its energy to the medium and thus appears redshifted. The latter will be discussed as a static universe nonexpansion alternative to the steady state model.

After World War II, three main trends emerged to explain universal expansion or a static model associated with redshift. The first was Zwicky's renewed tired light model by followers defending a static universe. The second was put forward by Fred Hoyle and consisted in a steady state model where minute amounts of matter are continuously created to compensate for the expansion and thus keep a constant cosmic density. The third was the big bang theory, proposed by Lemaître and developed by George Gamow, who introduced the concept of nucleosynthesis, and also by Ralph Alpher and Robert Herman, who predicted the microwave background radiation, whose discovery in 1964 strongly favored the big bang as the better of the two theories to explain the origin and evolution of the cosmos. From that time until now, the vast majority of investments, research, and publications have been directed at the development and defense of the big bang model since no viable alternative has been presented. It is my objective to present such an alternative.

1.2   Problems with Cosmological Models

Big bang, steady state, and static universe models bring some valuable input to the cosmological picture, but they are all riddled with flaws, including some unjustified infinite values [20]. The big bang model has too many mysterious and improbable solutions that have been added every time a problem comes up with observation. The second model, steady state, would require the incredible continuous creation, out of nothing, of new matter and just in the right proportions to fit the actual count as the universe expands. The tired light static universe alternative would require cosmic densities thousands of times above actual measurements to produce the observed redshift [16].

1.2.1   Big Bang Paradigm

When we extrapolate the big bang to its genesis at a theoretical 0 second, we are faced with infinite density and temperature at a finite time around 14 billion years ago [52]. The defenders of the model will assert that we cannot go beyond the Planck epoch at 10^{-43} second, when a blank occurs in physical laws that are no longer applicable, and so this qualifies as a first unexplained mystery. The period between the Planck epoch and 10^{-3} second is speculative and unresolved because energies exceed our human experimental capacities limited to about 10^{12} K, but this period also includes the mysterious inflationary epoch around 10^{-35} second, where the universe should have expanded exponentially in order to save the uniformity.

The model involves rapid growth and cooling with the passage from ultradense energy to quarks and neutrinos that will mute into matter–antimatter pairs with a positive segregation in favor of matter that will thereafter dominate and be represented by protons, neutrons, and electrons. Within a few minutes and at a temperature of about 1 billion K, lithium, deuterium, and a large fraction of helium were formed and had survived in a process called nucleosynthesis. According to the cosmological model, lithium nucleons cannot survive in most stars since they are transmuted to helium above a temperature of 2.4 million K, but colder brown and orange dwarf stars can keep their lithium fraction. Similarly, deuterium is destroyed very quickly in most stars by fusion into helium at a temperature as low as 1 million K but could be preserved from burning in colder bodies. Or could the two latter elements be generated in stars or hot gas clouds at about one billion K and be preserved by a rapid cooling process? The presence of about 25% helium-4 by mass or 8% by number of atoms in the universe is also accounted for by the nucleosynthesis model since stars cannot have produced this large amount of helium-4 from hydrogen fusion in the time frame since the big bang, but it could if a longer time period were available.

Around 379,000 years after the big bang, hydrogen and helium atoms are formed by the capture of electrons, and light is free to escape and supposed to form the cosmic microwave background (CMB) radiation, but since the velocity of light is far superior to expansion velocity, the question is, on what reflector will this escaping light bounce back to become the microwave background about 13.7 billion years later? How will it reach us? After such a big bang, we would expect to observe the equivalent of a large supernova remnant: a cosmos forming a huge spherical shell structure on which we would be living and with the original light gone far away from us forever. This is a questionable concept indeed.

The big bang proponents, assuming that space is like a stretchable material, invented a special space-expansion metric based on relativity called FLRW (Friedmann-Lemaître-Robertson-Walker) to explain space-expansion redshifts. The big bang model seems to be right on the general fact that cosmological redshifts from space expansion may be interpreted as a Doppler shift corresponding to a recessional velocity from which distances may be evaluated and a cosmic distance ladder may be built, although Hubble's law based on a linear relationship between velocities and distances is still debated for the nearby universe.

We will see later that cosmic redshift is the sum of a large number of specific cosmological space redshifts and blueshifts from expanding or contracting cells or bubbles across the cosmos and should be valid as a measure of distances on a large scale, but it does not imply a unique universal expansion. The Copernican principle stating that we are not in a privileged location near the center of the universe is an acceptable concept from past observational evidence. But on the contrary, the cosmological principle of homogeneity and isotropy, which is also established from observation, may be invalidated in a metric expansion model from a central big bang since its unique expanding shell is not an isotropic or homogenous spheroid.

An apparent mass excess was measured in the 1970s from a large number of rotation velocities computed from redshifts at the edge of galaxies or from clusters of galaxies [35]. This led to the conviction that about 90% of cosmic mass is in the form of nonbaryonic dark matter, which was never observed to this day. Moreover, an interpretation of accelerated expansion from IA supernova redshifts resulted in the hypothesis that about 70% of the total cosmic mass is in the form of exotic dark energy, for which there is absolutely no tangible proof but for which the cosmological constant was resurrected from Einstein's general theory of relativity. The problem is that this cosmological constant is 120 orders of magnitude smaller than the value expected from quantum gravity theory. We will demonstrate that these exotic dark energy and nonbaryonic dark matter can be excluded from a consistent cosmic model and that the finite universe is closed, thus excluding all flat or open hyperbolic models. There are also major problems in the big bang paradigm to explain large-scale structure formation and evolution since clear pictures from afar are generally not showing the predicted evolution.

1.2.2 Steady State and Static Universe Tired Light Models

In the steady state model, developed by Hoyle and associates in 1948, the universe is infinite in time: no beginning and no end. Instead of a one-shot, big-bang creation at the beginning, matter is steadily created to compensate for the decrease in density due to expansion, and thus, the cosmological principle of homogeneity and isotropy is perfectly respected. The amount of new matter needed is about one hydrogen atom per cubic meter per billion years, and also needed is a mass of exotic dark matter five times that of baryonic matter. To this end, Hoyle proposed a "creation field" (called C-field) with negative pressure to act against gravity and produce the measured expansion. The model was later modified by replacing continuous creation with discrete minipockets of creation.

A first major problem is that with infinite time, we expect infinite accumulation of by-products from processes like nuclear fusion, and a second major problem stems from the infinite material creation in infinite space expansion, which does not make sense in a cosmos filled with finite units of material. Since no mechanism for primordial nucleosynthesis of lithium, deuterium, and most helium is known in stars, this model would imply the unlikely continuous creation in the right proportion of those primordial elements. Also in this model, the microwave background radiation is explained as light from old stars that is scattered on dust. The problem is that grains of dust vary extremely in size, temperature, and scattering properties, so we should not find the smooth microwave background that is observed as a fact.

Zwicky's tired light model is the subject of renewed interest and discussions in recent years [27]. Among others, Paul Marmet wrote many papers where he explains that semiclassical electrodynamics and quantum electrodynamics theories consider all interactions between photons and atoms as inelastic, which imply photons losing some energy. The basic idea is that photons traveling across the cosmos are not subject to a cosmological Doppler redshift but are gradually losing an infinitesimal part of their energy, about 10^{-13}, every time they interact with an atom or molecule and that the energy lost is proportional to measured redshift. But the problem is that an average density of about 10,000 hydrogen atoms/m^3 is necessary to produce this effect. This density represents about fifty thousand times the density of the cosmos as measured from star and gas density and from the primordial nucleosynthesis theory. It is so large that the universe would rapidly collapse in parts and in totality. Also, the Tired Light model does not predict the observed gradually increasing width factor and related decay time delay for the redshifted light curves of supernovae located farther and farther away.

1.2.3   Occam's Razor and Proposed Model

William of Occam was an English Franciscan friar who lived in the 14th century and to whom is attributed the principle of using the razor to shave off extraneous assumptions and to use the smallest number of hypotheses to solve a problem. It is not a proven law or theory but an approach to science and knowledge where simplifying usually brings a better result, but not always. It is a general guiding rule in the development of models in physical sciences that will be applied heuristically to the proposed model in this writing.

The basic geometry of the cosmos will be introduced in the shape of spheres and bubbles from which the cosmic mass distribution will be established for our spheroidal universe. Cosmological redshift will be presented as associated with every local bubble expansion and not as a total universal expansion. We will follow some of the light travels with refraction and interaction across cosmic clouds, filaments, and other medium, all the way to various backgrounds, including the cosmic microwave background.

The origin of every bubble expansion will be rooted in active galactic nucleus and quasar contraction followed by a maxibang as the source of local nucleosynthesis and of a new, expanding shell of material that will locally create the next generation of stars, galaxies, and clusters. The apparent mass excess in galaxies and clusters of galaxies will be explained by real and apparent velocities related to the new expansion dynamics and to non-Newtonian centers of gravitational attraction. So exotic dark matter and dark energy will no longer be needed since no mass excess is present. Finally, the four forces or bonds of energy in nature will be presented as the gravitational, the black, the electronuclear, and the biochemical.

2. MASS DISTRIBUTION AND DENSITY IN COSMOS

After a brief look at efficient minimum distances in a two-dimensional surface pattern and a volumetric review of solid spheres in terms of a geometrical arrangement in two or three dimensions, we will see the spatial distribution of expanding bubbles, walls, and junctions between bubbles. Mass and density of stars, clouds, filamentary structures, galaxies, and clusters will be presented in order to establish an average cosmic density, an average radius, and the total mass of a finite cosmic model.

2.1 Spheres

A sphere represents a basic geometrical shape in nature for which we find the smallest minimum surface for a given volume. Spheres are found at all scales and levels of complexity, from the atomic and molecular scale to complex organisms like cells, and they are most important in astronomy at small and large cosmic scales.

2.1.1 Free Space in Two Dimensions

Before discussing spheres, let us have a look at some efficient two-dimensional surfaces. What is the shortest trail or minimum road system between four points forming a square whose side has a length of one unit in two-dimensions (2-D)? Looking at figure 2.1, we see that distance AB = BD = DC = CA = 1, so the sum of the four sides is equal to 4. It is proven that the network, shown in the figure with angles of 120° between every two segments around points E and F or triple-point junctions, yields the minimum value of $1 + \sqrt{3}$, or 2.73 units, when we add all the segments. This is much less than for the square where a sum of 4 was found. In fact, it is about 4% less than the next minimum value of $2\sqrt{2}$, or 2.83, found when we directly connect A with D and B with C as a crossroad system.

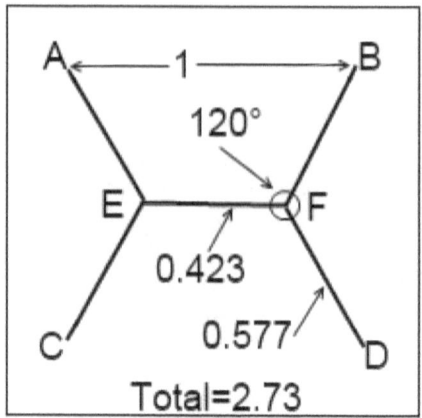

Figure 2.1 Shortest or minimum trail system between four points

This pattern of minimum length on a surface, also shown in figure 2.2 and associated with angles of 120° around triple-point junctions, is frequently found in nature and represents a way to use a minimum amount of energy. For example, bees make their honeycomb in this fashion with a minimum of wax. But how they measure the 120° angles as they gradually build the comb from top to bottom is still an unanswered question. It is likely some kind of innate behavior, like the dance they perform to indicate distances and directions to new flower fields to fellow workers. Also, triple points or hexagonal patterns are found in frost and snow crystals since they are made up of water molecules with angles close to 120° (actually 104.45°).

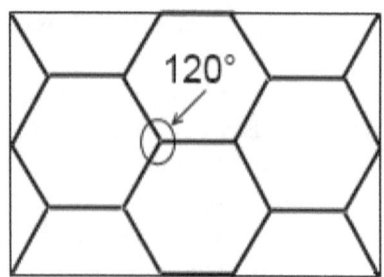

Figure 2.2 Hexagonal or honeycomb pattern

If we pile up circles of equal size, like disks, for example, on top of one another in a square or rectangular arrangement, we end up with something looking like figure 2.3, shown in two-dimensions (2-D). Using a radius of one unit, we find a surface of π, or 3.1416, for each circle enclosed in a square that has a surface of 4, so the circular surface is only 79% of the total and the empty surface is about $^1/_5$ of total at 21%. But this is not an efficient way to occupy the space available with identical circles and to minimize the empty area.

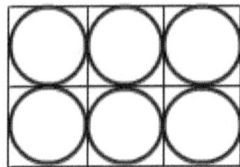

Figure 2.3 Circles in a square arrangement

For a two-dimensional surface array, figure 2.4 is showing that the hexagonal packing, where a circle is tightly surrounded by six equal circles, will leave much less space between circles. This pattern can be repeated on and on to cover any surface. J. L. Lagrange proved in 1773 that this lattice arrangement has the highest density, corresponding to $\pi/\sqrt{12}$, which is about 91% of the surface. This leaves only 9% of the surface for all empty areas located between any three circles and corresponding to those acute triangles with three angles less than 60° each for a sum of angles less than 180°.

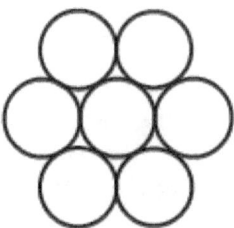

Figure 2.4 Hexagonal close packing surface in 2-D

When we compare with figure 2.2, we realize that the same horizontal and diagonal efficient lattice arrangement is present: one showing hexagons and the other showing circles.

2.1.2 Free Space in Three Dimensions (3-D) and Kepler Conjecture

When we look at three-dimensional (3-D) space instead of 2-D, we discover different space fillings. Imagine figure 2.3 as a 3-D drawing of spheres in a boxy or cubic arrangement. Then, we would find that each sphere with a radius of one unit represents a volume of $4\pi/3$, or 4.19, which amounts to 52% of 8, the total cubic volume, and that the empty or void space around the sphere represents 48%, which is almost half the total volume. But we will see that the Kepler conjecture offers a more efficient packing of spheres with minimum free space in between. It corresponds to the way spherical fruits like oranges are usually and naturally stacked up in a grocery store, starting at the bottom with a compact hexagonal lattice as on figure 2.4.

The Kepler conjecture has only been proven recently in 1998 by Thomas Hales. It is a 400-year-old geometric conjecture stated by astronomer Johannes Kepler, who wrote in 1611 about the packing of spheres in 3-D space [64]. He asserted that we cannot fill space with a greater density than these two equivalent arrangements of equally sized spheres: the first being a face-centered cubic close packing and the second being called the hexagonal close packing as shown in figure 2.5. Note that three spheres have been separated from both the top and the bottom of the two arrangements to give a better view and that the six central spheres are placed in the same hexagonal order as the 2-D example of figure 2.4. The difference between the face-centered and the hexagonal close packing is the rotation of the top three spheres with an angle of 60°, which gives a third layer to the arrangement. We will see later that this rotation has an important consequence in the arrangement of free space for galaxies in a cosmological model. The face-centered cubic arrangement is named from its 3-D lattices forming cubes, where each face is made up of five spheres: like letter X, with four spheres in the corners and one at the center. The latter has three different layers, or planes, that repeat themselves. But the hexagonal arrangement has only two different layers repeating themselves on top of one another, like $ABAB$.

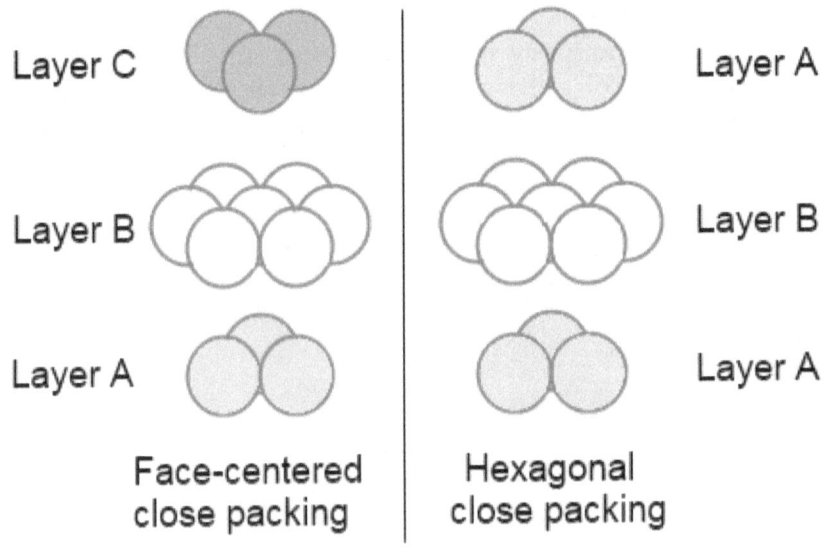

Layer C

Layer B

Layer A

Face-centered close packing

Layer A

Layer B

Layer A

Hexagonal close packing

Figure 2.5 Face-centered and hexagonal close packing

The close packing sphere density of both arrangements is $\pi/\sqrt{18}$, which is equal to 0.74, or 74%, of the total volume, so maximum empty space left to be filled with filamentary structures and clusters of galaxies is 26% if, of course, the cosmos could correspond to such a perfect and rigid model. We observe the following facts for the spheres involved in these arrangements:

- Each sphere is in contact with twelve other spheres from which a dodecahedron may be formed.
- A triangular space with three acute angles less than 60° is present at the junction of any three spheres as shown on figure 2.4.
- An opened tetrahedral volume is present at the junction of any four spheres, so each sphere may be attributed ¼ of that space.
- Each sphere shares eight tetrahedrons with other spheres.
- So eight tetrahedrons shared by four spheres each gives 8/4 = 2 tetrahedrons per sphere, which could correspond to two secondary clusters of galaxies per sphere on a cosmic scale.
- The distance between two tetrahedral volumes is equal to the length of the filament or duct formed by the triangular space between any three spheres and measures approximately 15% more than the spherical radius, or less if compaction is present.
- There are also six hexahedral openings associated with each sphere. Three hexahedrons are shown on figure 2.6.

- Since each hexahedron is surrounded by six spheres, this yields one hexahedron per sphere: these are the primary large openings where gas and galaxies will gather and form large clusters of about one thousand galaxies plus associated gas that will gradually be compressed and produce the future bangs.

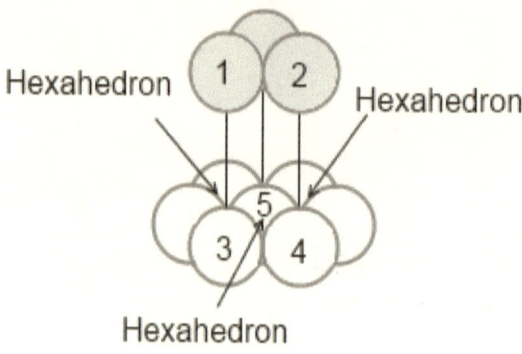

Figure 2.6 Three hexahedrons in the hexagonal close packing

On figure 2.6, we indicate with three vertical lines the three tetrahedral cavities that will be covered by the bottoms of the three upper spheres, and we also show where the three top hexahedrons are located in a hexagonal close packing arrangement. The three bottom hexahedrons are not shown but are located directly below the three top ones, for a total of six around sphere number 5. Only five of the six spheres forming the front hexahedral space are indicated since the sixth one, up front, would hide the others. An important relationship to notice is that in the face-centered cubic close packing, the top hexahedrons are rotated with an angle of 60°, so they are not located directly on top of the bottom ones, but they are staggered and they form a different plane.

2.1.3   Space Reduction between Gaseous Spheres

Four spheres in a close packing contact are shown surrounding an opened tetrahedral volume on figure 2.7 where a cluster could be nested with filaments coming out from the four triangular openings between any three spheres.

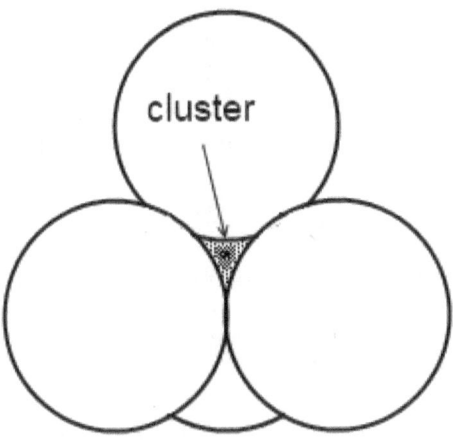

Figure 2.7 Four spheres surrounding a secondary cluster of galaxies

But we are not dealing with rigid spheres in intergalactic and intercluster space. On the contrary, we find decompressing and cooling gaseous walls with filamentary structures surrounding large, extremely low-density spherical and ellipsoidal structures with flexible shapes.

2.2   Bubbles

We will now go from rigid spheres to more flexible spherical forms and look at the shape and special characteristics of some bubbles, especially when we combine two, three, or four bubbles to form common walls. Soap bubbles represent physical examples of complex mathematical problems related to minimal surfaces [75].

2.2.1   Surfaces for Two Bubbles

As shown on figure 2.8, when two bubbles of equal size join together, they form a common wall with a flat surface because water mixed with soap, for example, decreases the water surface tension and thus naturally helps to keep the liquid film from breaking and minimizes the areas for the enclosed volumes. So this simple double-bubble model contains three different surfaces made up of two partial spherical minimum surfaces for the enclosed volumes and a flat disk forming the common wall.

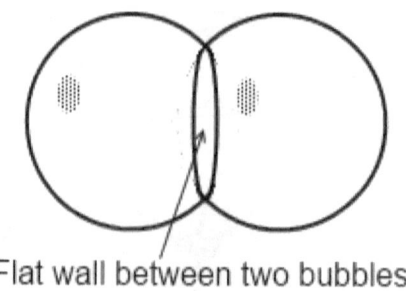

Flat wall between two bubbles

Figure 2.8 Minimum surfaces between two bubbles

It was only in 1995 that Joel Hass and Roger Schlafly announced a proof of this conjecture, stating that three spherical surfaces are formed when two bubbles of equal volume coalesce, the central surface being flat in this special case with a radius equal to $\sqrt{3}R/2$, or about 87% of bubble radius R.

On the other hand, the proof that two merged soap bubbles of unequal size, as on figure 2.9, yield the best way of enclosing two different volumes of air with the minimum surface area is still open for proof.

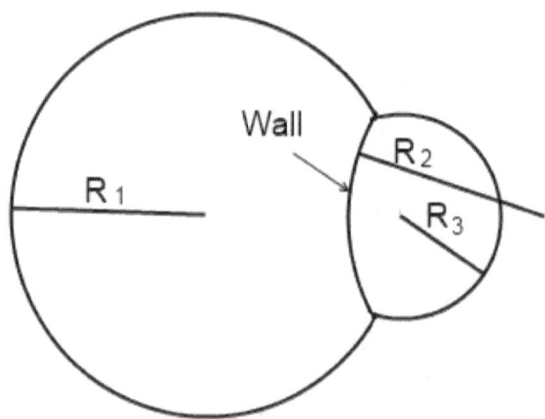

Figure 2.9 Small bubble merging with a larger one

The greater surface tension and, thus, pressure in the smallest bubble produces the incurving inside the larger bubble. But nevertheless, if the double-bubble conjecture applies to this asymmetric case as shown on empirical ground, then the radiuses of the three spherical surfaces must satisfy the following relationship in which the radius of the two bubbles are R_1 and R_3 and the radius of the common surface is R_2, yielding this equation:

$$1/R_1 = 1/R_3 + 1/R_2 \qquad (2.1)$$

For example, if $R_1 = 1$ and $R_3 = 1/2$, then $R_2 = -1$.

2.2.2 Junction of Three Bubbles

On figure 2.10, we are showing a 2-D sketch representing three coalescing equal-volume and equal-surface bubbles that are each separated from their spherical neighbors by two flat surfaces.

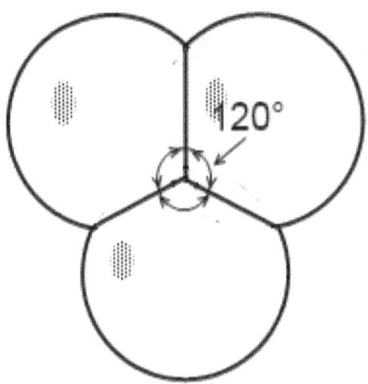

Figure 2.10 Junction of three bubbles

These flattened surfaces form three angles of 120° around the central line joining all three surfaces.

2.2.3 Tetrahedral Space between Four Bubbles

The first scientific experiments with soap bubbles were initiated by Joseph Plateau in the 19th century. He was the first to demonstrate physical solutions to the minimization problems, using liquid soap with a variety of geometrical shapes and polygons. This experimenter stated three geometrical conditions always met by soap films bounding any frame:

- The three smooth surfaces of a film intersect along a line, as on figures 2.10 and 2.11.
- The angle between any two surfaces at any point along the line of intersection of three surfaces is 120°, as on figures 2.10 and 2.11.
- Four of the lines, each formed by the intersection of three surfaces, meet at a single point, and the angle between any pair of adjacent lines is 109.28°, as shown in figure 2.11.

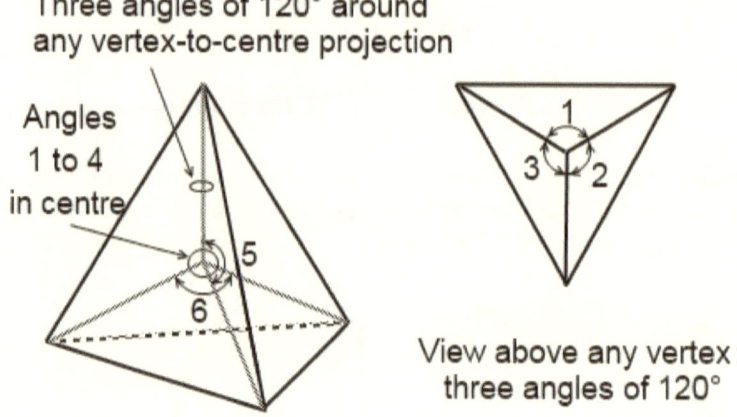

Three angles of 120° around
any vertex-to-centre projection

Angles
1 to 4
in centre

View above any vertex:
three angles of 120°

Four bubbles meeting at centre
forming six angles of 109.47°

Figure 2.11 Angles and vertices at the junction of four bubbles

We can view this meeting of four bubbles at the center of the tetrahedron if we visualize the four spherical portions of the bubbles outside each face of the tetrahedron as seen on figure 2.12 with the four vertices indicated. The dotted lines in the background show the fourth bubble and the base of the tetrahedron joining the vertices.

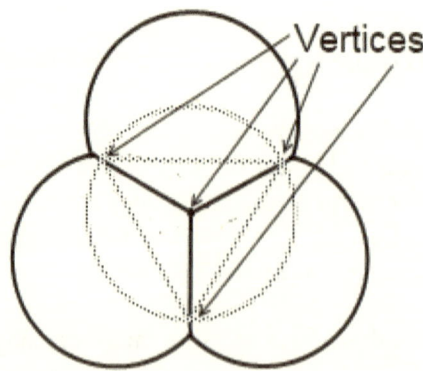

Vertices

Figure 2.12 View above one vertex and facing three of the four bubbles

This is the same view above one vertex as on the right of figure 2.11, where the spherical part of the bubbles is absent. Note that a wall is present between any two bubbles for a total of six separate walls.

2.3   Density and Mass Distribution

The permanent and finite cosmos is filled with energy and knots of energy observable from a variety of qualitative and quantitative characteristics. Now, let us make an assessment of densities and associated masses at a cosmological scale, including stars and their satellites, filamentary gas structures and galaxies, clusters and their nested hot gas, and low-density bubbles or voids with their sparse galactic and gaseous inhabitants. Note that although *void* is a good term to represent the extremely low density of a large cosmic volume, the term *bubble* will be used most often for its more dynamic and punchy significance [43].

2.3.1   Density of Stars and Their Satellites

Here on earth, we live in a high-density medium where the three phases identified as gas, liquid, and solid, added to a touch of plasma, exchange their energy in a relative equilibrium. Air, for example, with the lowest density of the three phases, is a mixture of gas with a density of 1.2 kg/m^3 at sea level for a temperature of 21°C. The average earth density is 5,515 kg/m^3, similar to the density of other rocky planets and satellites in the solar system. Our sun, like other stars, is mainly composed of plasma, where atoms are stripped of their electrons, more so toward the radiating center, where densities and temperatures reach atomic fusion levels. Its average density of 1,408 kg/m^3 is only 41% more than water density of 1,000 kg/m^3, but varies enormously between the corona at 1.0×10^{-12} kg/m^3 and fusion center at about 1.62×10^5 kg/m^3. Hydrogen nuclear fusion stars have a range of masses between about 8.5% of solar mass for the subdwarfs and more than 100 times the solar mass for the hypergiants, with proportional density variations. If we exclude hypothetical massive singularities and black holes, the champions of density are the gravitationally collapsed massive stars called neutron stars, with a compacted mass of 1.35 to 2.0 times the solar mass in a sphere with a radius of about 12 km and a staggering density up to 6.0×10^{17} kg/m^3, which is twice the atomic nucleus density.

2.3.2   Galaxy, Cluster, and Bubble Densities

If we compute the average density for the equivalent of 130 billion solar masses in our Milky Way galaxy, using a radius of 65,000 light-years, expressed hereafter with the symbol l-y, and a thickness of 1,000 l-y, we find a relatively low value of about 2.8×10^{-20} kg/m^3, which may also be expressed more conveniently as 17 million M_p/m^3, where M_p means "proton mass" or about the mass of one hydrogen atom. Then, the next step up is represented by groups of about forty galaxies where a radius of 5 million l-y in a spherical volume yields a much lower density value of 3.34×10^{-27} kg/m^3, which is equivalent to 2.0 M_p/m^3. Clusters of galaxies are the largest known massive structures in the universe: one cluster has an average radius of about 13 million l-y and contains an average of five hundred galaxies, which gives, in a spherical volume, an average density of 2.0×10^{-26} kg/m^3 or 12 M_p/m^3. On the other hand, spheroidal or ellipsoidal expanding bubbles are the largest low-density structures, much larger than clusters of galaxies: with an average radius of 91 million l-y, their average density of 3.9×10^{-29} kg/m^3, or 0.023 M_p/m^3, is about ten times less than the universal average of 0.21 M_p/m^3. The presence of 5% of the galaxies in residual walls and filamentary structures within the bubbles has been included in the balance sheet.

2.3.3   Average Cosmic Density

About 13.8 billion years ago, our vast and entire local bubble, estimated to have a present radius of 100 million l-y, including all related galaxies, neutrinos, and electromagnetic energies, was compressed into an extremely small volume and bounced back on a gravity wall. Like the phoenix reborn from its own ashes, this maxibang event represents the birth of our bubble out of the ashes of previous galaxies and other captured and compacted masses and energies. This birth, after a gravitational compaction and bounce back, marks the beginning of our local space and time.

A chronology of the first instance into expansion, similar to the one described by the big bang model, is expected for every maxibang in the cosmos. This includes that period of expansion between about 3 and 20 minutes when nucleosynthesis occurred at a temperature of about 1 billion Kelvin (K) and a density, like air on earth, of about 1.2 kg/m^3 or 7.2×10^{26} M_p/m^3. As expansion and cooling took place, protons survived the antiprotons' mass destruction and formed hydrogen nuclei. Figure 2.13 presents a summary of some published data on the average abundance of helium-4, helium-3, deuterium, and lithium-7. These ratios are calculated, as shown on the graphs, to be 12.5 hydrogen atoms for each helium-4, 30,000 hydrogen for each deuterium atom, 90,000 hydrogen for each helium-3, and 5 billion hydrogen atoms for each lithium-7. The result indicates a rough agreement among all four values in which deuterium is a good match and helium-4 is an acceptable match, but there are important uncertainties for helium-3 and lithium-7. Since stars cannot produce this large amount of helium-4 in the time frame of our bubble, it sets a constraint on the model [51].

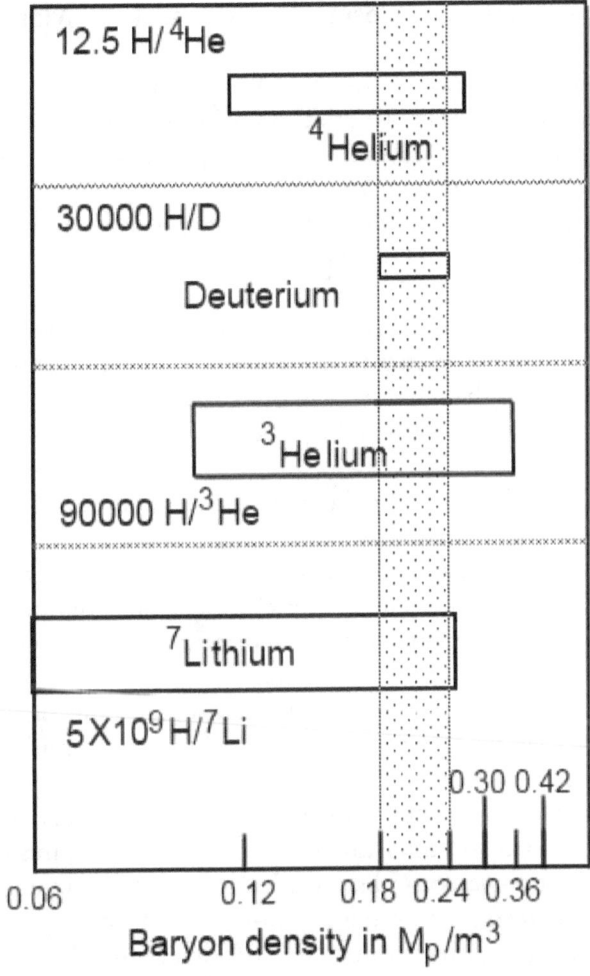

Figure 2.13 Baryon density in M_p/m^3 from nucleosynthesis

All these abundances are related to a single parameter, which is the ratio of photons to baryons calculated from cosmic microwave background fluctuations. So all four nucleosynthesis elements could have a common photon-to-baryon ratio, shown by the dotted vertical strip centered on about 1.65×10^9 photons per baryon and ranging in baryon density from 0.18 to 0.24 M_p/m^3 from which we will use the average of 0.21 M_p/m^3 for the entire cosmos. Evaluations of neutrino and electromagnetic mass density input to the total cosmic count generally indicate negligible values.

2.3.4 Average Cosmic Radius and Total Mass

From Kepler's and Newton's equations, we can find the velocity of a satellite with a negligible mass in a circular orbit around a more massive and spherically symmetric or point-like central body, using the well-known equation

$$V = \sqrt{GM/R} \tag{2.2}$$

where V is the circular velocity, G is the gravitational or Cavendish constant, M is the mass of the massive body, and R is the circular radius of the orbiting satellite. This equation yields density and radius for the middle curve in figure 2.14, where the horizontal scale runs from a radius of 10 to 200 billion l-y and the vertical scale covers densities from 0.01 to 100 M_p/m^3. But if we multiply by two the term under the square root in this circular orbit equation, we obtain the escape velocity or Schwarzschild equation

$$V = \sqrt{2GM/R} \tag{2.3}$$

from which the density and radius are extracted and shown as the bottom curve on figure 2.14.

However, in a closed and finite permanent spherical universe, we do not want any particle or any kind of energy to escape, not even photons, so the Schwarzschild equation is unacceptable and rejected since some of the mass energy would definitely escape in this case.

On the other hand, equation 2.2 is telling us that some energy or matter may also escape even though it would be limited to a circular orbit around the central mass. The problem is, when we use the fastest velocity, which is the velocity of light, we can expect the accumulation of photons and light particles overcrowding the circular orbits around the central mass, so this option must also be rejected since we don't observe a leaking and an evaporating cosmos. Thus, in order to keep a lid on the cosmos and to avoid any escape of mass energy from multipole irregularities, we will divide the right-hand side of equation 2.2 by two to obtain the following modified equation:

$$V = \sqrt{GM/2R} \tag{2.4}.$$

This is the modified equation that is applied in figure 2.14 for the top curve with the triangular symbols. The selected density of 0.21 M_p/m^3 is reached, as indicated by the arrow at a radius of 143 billion l-y. This cosmic radius is about ten times more than big bang estimates, but the density of 0.21 M_p/m^3 for ordinary baryonic matter is the average predicted by the big bang model.

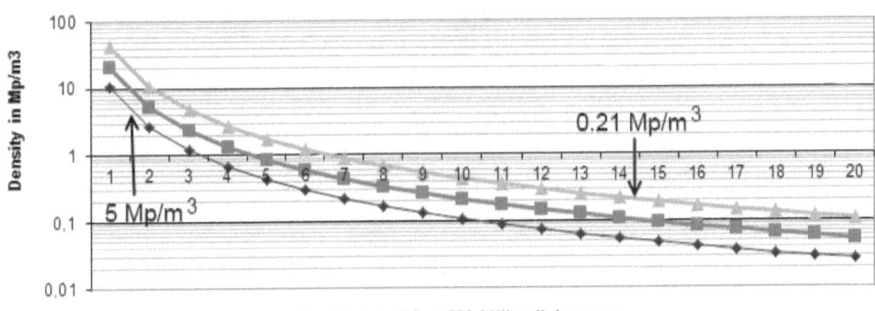

Figure 2.14 Distribution of density and radius for three models

Also, note that for earlier big bang models, before the reintroduction of the cosmological constant, a value of 5 M_p/m^3 is shown with an arrow pointed at the Schwarzschild curve, which indicates the approximate value of the critical density that makes the cosmos stands on the edge between expansion and contraction. But more recent big bang models of the Friedmann universe have added hypothetical dark matter, whereby the actual density is equal to the critical one, and also hypothetical dark energy to explain that the universe is not expected to contract but seems to accelerate its expansion so that the latter should last forever and gradually end up in a dissociated and completely destroyed, cold, dead cosmos when all the electronuclear energy is dispersed and consumed [56].

2.3.5   Finite and Permanent Cosmos

What is the topology of the cosmos? With a known radius and a permanent spheroidal model, what does the cosmos look like? Figure 2.15 is a 2-D representation of the cosmic multipole spheroid filled with bubbles and clusters and where particles and photons go grazing near the edge without ever escaping and without reaching a circular orbit either. Particles will fall back more quickly than light will and in proportion with their respective speed near the edge, as more massive particles, like atomic nuclei, need more energy to be accelerated close to the speed of light. Of course, time, mass and momentum relativistic effects and also some polarization may be present. Bubbles and clusters are in equilibrium between expansion and contraction: as bubbles expand in the universe, they keep an outward pressure that counteracts gravitational attraction between clusters. Whenever the combination of bubble pressure and gravity forms a critical massive knot of material, such as a cluster of galaxies with a mass of about one thousand galaxies plus enclosed gas, this same knot will collapse on itself before bouncing back and expanding as a new bubble in a new cycle. This way, the cosmic principle of homogeneity and isotropy is respected at large scales, and there is no privileged center in the universe. The cosmos is a self-regulated cyclic system of growth and contraction like the cells, in a body with a constant volume, are born and grow to replace dead cells. It may also be compared to the renewed bubbles in a flute of champagne.

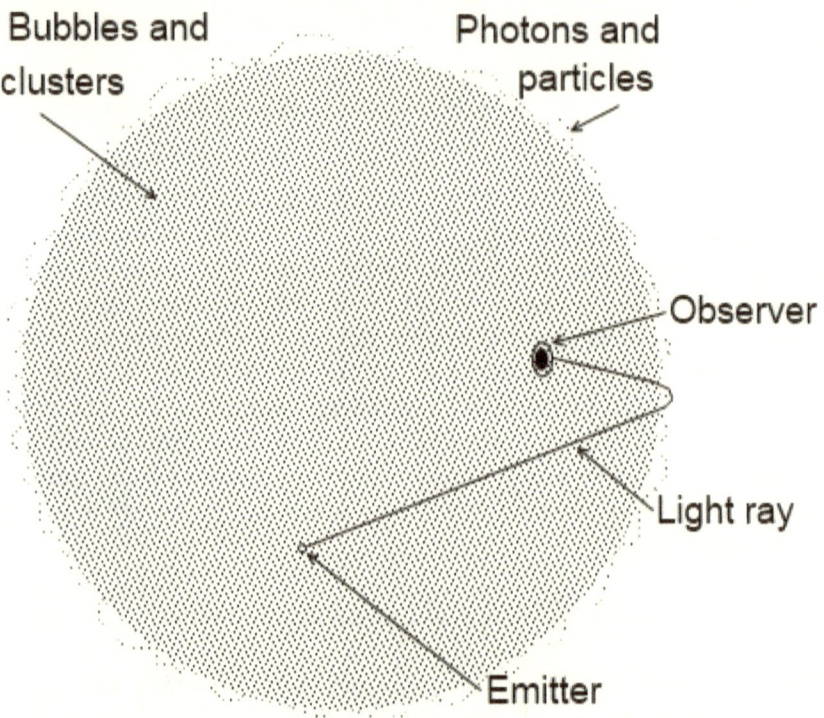

Figure 2.15 Spheroidal bright hole cosmos

One of many possible electromagnetic wave or light trajectory between an emitter and an observer is also shown on the figure. We want to emphasize the fact that photons are always deviated when traveling across the cosmos by refractions and lensing effects from the gravitational pull of a multitude of massive bodies and especially by the cosmic mass itself. Moreover, light may be absorbed and reemitted as reflections, refractions, or scatterings. But in about 90% of the cosmos, light reaches its maximum velocity, c, and is little hampered when it travels across vast, expanding bubble volumes in which extremely low densities and weak fields prevail. So these bubbles may be considered as slightly freer nonreactive corridors and fast tracks for light across the cosmos.

Selected cosmic parameters:			
Physical parameter	Number	Units	Remarks
distance: 1 parsec (pc) =	3.26	light-years (l-y)	1 kiloparsec= 1 kpc
distance: 1 l-y =	9.47E+15	meters (m)	E means "exponent" in column 2
cosmic density	0.21	Mp/m³	proton mass/cubic meter
average cosmic radius	1.43E+11	l-y	light-years
average cosmic radius	1.3542E+27	m	meters
cosmic volume	1.0403E+82	m³	cubic meters
cosmic mass	2.1846E+81	Mp	proton masses
proton mass	1.67E–27	kg	kilogram
cosmic mass	3.65E+54	kg	
bubbles mass	3.65E+53	kg	10% of total
interbubble mass	3.28E+54	kg	90% of total
bubbles volume	9.3624E+81	m³	90% of total
bubbles mass	2.18E+80	Mp	proton masses
bubble density	0.02333	Mp/m³	proton mass/cubic meter
neutrino density	1.15E–03	Mp/m³	1.00E+8/m³
neutrino mass	2.00E+52	kg	0.001 to 0.01 of total cosmic mass
electromagnetic density	3.00E–04	Mp/m³	proton mass/cubic meter
electromagnetic mass	3.65E+49	kg	CMB = 6.0E-5 of total cosmic mass

Table 2.1 Selected bubble and interbubble cosmic parameters

In the first column of table 2.1, we find the physical parameters being evaluated, with the quantitative result in the second column, followed by the corresponding unit in the third column and comments, as applicable, in the last column. The value of cosmic density is determined with the baryon density from nucleosynthesis as discussed above. The mass of bubbles is tentatively evaluated as 10% of the total cosmic mass, but it could vary around this value, some authors going as low as 5%. Expressing cosmic-scale densities in term of proton masses per cubic meter is a practical unit since hydrogen (H) is the most basic and most spread element in the cosmos, with a single proton having a mass 1,836 times larger than its electron in the neutral configuration.

Neutrino data are still under experimental scrutiny since the mass is only estimated in a bracket between 0.001 and 0.01 of total cosmic mass, suggesting an influence less than 1% of the total. It is evaluated that the cosmic microwave background, which contributes the most to the electromagnetic density of the cosmos, represents a scant value less than ½ of a million of the total cosmic mass.

3. ELECTROMAGNETIC WAVES, DOPPLER EFFECT, AND REFRACTION

What is this almost ubiquitous substance without which there would be no astronomical observations and without which astrophysicists could not apply or extrapolate the laws of physics? Of course, it is light and its electromagnetic wave family, which also has a double nature expressed as photon particles bearing energy proportional to frequency and wavelength [60].

Class	Wavelength	Frequency	Energy	
Gamma rays	1 pm	300 EHz	1.24 MeV	
X rays Ultraviolet	1 nm	300 PHz	1.24 keV	Visible
	$1\,\mu m$	300 THz	1.24 eV	light
Infrared Microwave	1 mm	300 GHz	1.24 meV	
Television-FM	1 m	300 MHz	$1.24\,\mu eV$	
AM radio	1 km	300 kHz	1.24 neV	
EM voice Extemely low	1 Mm	300 Hz	1.24 peV	
Planck's constant	300 Mm	1 Hz	4.14 feV	

Table 3.1 Electromagnetic wave summary

The symbols in table 3.1 are as follows: *m* means "meter" when standing alone or following another symbol; *Hz* means "Hertz" or "cycle per second"; *eV* means "electron volt." The first symbol preceding *m*, *Hz*, or *V* gives in decreasing order the size in multiples of 1,000, except for the last one related to Planck's constant:

E means "exa" and multiplied by 10^{18}

P means "peta" and multiplied by 10^{15}

T means "tera" and multiplied by 10^{12}

G means "giga" and multiplied by 10^{9}

M means "mega" and multiplied by 10^{6}

k means "kilo" and multiplied by 10^{3}

none means "unity" and multiplied by $10^{0} = 1$ (no symbol present)

m means "milli" and multiplied by 10^{-3} (not to be confused with *meter*)

μ means "micro" and multiplied by 10^{-6}

n means "nano" and multiplied by 10^{-9}

p means "pico" and multiplied by 10^{-12}

f means "femto" and multiplied by 10^{-15}

a means "atto" and multiplied by 10^{-18}

But in astronomy and astrophysics, scientists often exceed these values both at the small—and large-scale ends.

The first column of table 3.1 shows electromagnetic waves (EM waves) from gamma rays at the high-energy end to Planck's constant at the lower limit and where energy *E* is the difference between two quantum states and equal to the product of Planck's constant *h* by the frequency *f*:

$$E = hf \tag{3.1}$$

This is called the Planck formula, in which frequency *f* is also proportional to the speed *v* of the wave divided by wavelength λ:

$$f = v/\lambda \tag{3.2}$$

where *v* may be replaced by symbol *c* for the speed of EM waves.

A thin line across the chart indicates the position of visible light, which is only a tiny portion of the electromagnetic waves.

3.1 Doppler Effect in Daily Life

After a review of Doppler's effect related to movements of source and/or receptor, we will look at refraction in a medium and present the prism as an example [59]. Astrophysics would not be the same if the Doppler effect associated with light was not present in the cosmic realm. But as earthly observers, we experience the effects of Doppler shift in our daily lives, mostly with sounds. For example, when a train moves toward us, the apparent sound frequency is higher than emitted frequency, and when it moves away, frequency is lower than emitted frequency.

3.1.1 Doppler Shift Related to Source and Observer

The emitter is the source of the wave disturbance, and the receptor represents the observer in the bottom sketch of figure 3.1, where four complete cycles are represented. When a sound or light wave is emitted by a source, it travels across the surrounding medium—or low-density field while it is apparently stretched or contracted and until it reacts by absorption, refraction, reflection, or scattering.

Note that sound waves are carried by the medium in which they travel, like the wind, for example, for which the velocity of the wind adds up or is subtracted from the velocity of the sound wave, depending on direction. Light or EM waves, on the other hand, react with the surrounding medium in relation to its density, but the Michelson-Morley experiment demonstrated that light is not carried by a medium such as ether, so the speed of light is considered as a constant for any moving reference frame such as earth.

Light travels at a maximum official speed of $c = 299{,}792{,}458$ m/s in the low-density medium of the solar neighborhood and should travel at a very slightly higher speed in intergalactic extremely low-density medium. In practice, we will use a rounded value of $c = 300{,}000{,}000$ m/s in our calculations. At our scale, we are not sensitive to Doppler light shifts on earth because the speed of light is so much larger (about 1 to 10 million times more) than our familiar speeds, such as a car at 30 m/s and even sound at a speed of only 330 m/s. But at our human scale, Doppler shifts from sound waves transmitted in air as a medium are the most familiar and easy to measure since the car speed at 30 m/s is a substantial fraction or about 9% of the speed of sound at 330 m/s.

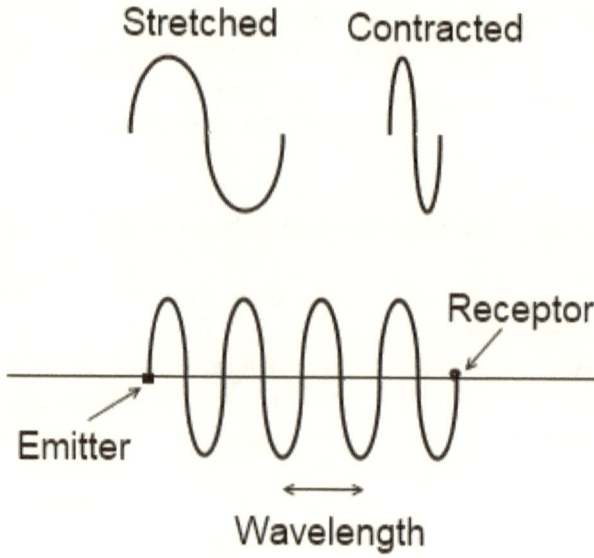

Figure 3.1 Doppler shift related to fixed or moving emitter and receptor

When both emitter and receptor stand still in a homogeneous medium where friction and other energy losses are neglected, then the wave is transmitted with a constant frequency, wavelength, and amplitude, and no Doppler effect is present. If either the emitter or the receptor or both are moving toward each other or away from each other, then Doppler shifts may be recorded. Whenever the net displacement between emitter and receptor brings them closer, then frequency, apparently increases and wavelength is contracted, as shown on the upper right of figure 3.1, and this translates into a blueshift if light is involved. On the contrary, if the displacement takes emitter and receptor apart, then frequency apparently decreases and wavelength is stretched, as shown on the upper left of the figure, and this corresponds to a redshift if light is involved.

3.1.2 Frequency Shift When Receptor Is Moving away

The Doppler effect when the observer or receptor is moving away is illustrated in figure 3.2, using only one cycle in 1 second (1 Hertz) to visualize more easily the phenomenon. For sound waves, we would multiply the number of Hertz by 20 to 20,000 for the range of human audition.

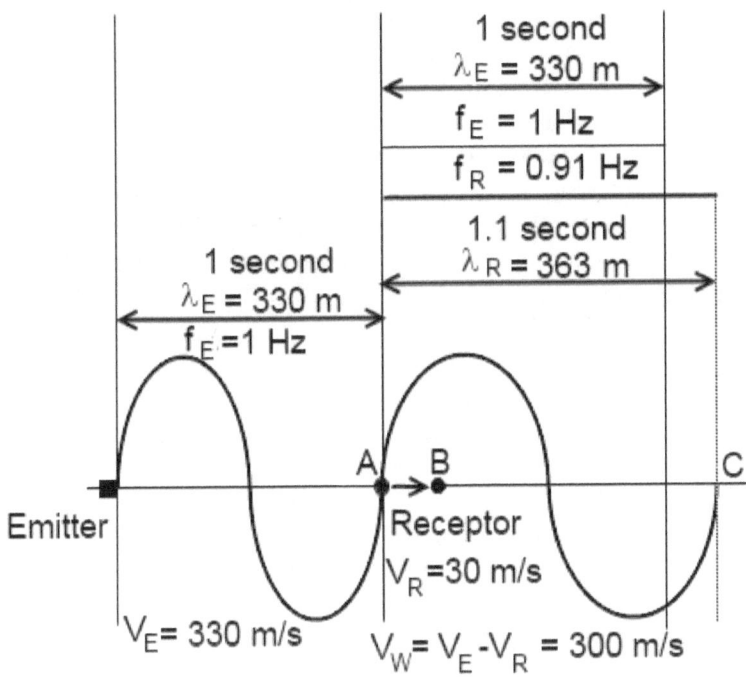

Figure 3.2 Doppler effect when receptor is moving away

The vibration is generated by the emitter and travels in the air as a medium at a speed of 330 m/s and a frequency of 1 Hertz. From equation 3.2, we find a wavelength of 330 m, and since the medium does not change, the speed, frequency, and wavelength of the vibration are the same, second after second, away from the emitter and independently from any observer, as shown after 2 seconds on the upper right of the figure.

But if after 1 second an observer is moving from point A to point B at a speed of 30 m/s and away from the source emitting the signal, then the signal has to catch up with him, and he will measure a lower wave speed V_W of 300 m/s for that signal, which is equal to the difference between emitted speed V_E and receiver speed V_R, so we have

$$V_W = V_E - V_R \qquad (3.3).$$

Combining equations 3.2 and 3.3, we find $f_R = V_W/\lambda_E = 0.91$ Hz, which is the observed lower frequency, representing less than a complete cycle in 1 second. But to receive a complete wavelength or a full cycle of 330 m from the emitter, our moving observer needs more time. He receives the signal at a speed of 300 m/s, which means that the time needed is $t = \lambda_E/V_W$ and equal to 1.1 second and corresponds to an observed wavelength of $\lambda_R = \lambda_E t = 363$ m.

Note that the energy of the emitted wave is basically conserved since it is stretched out and received over a longer period for the observer. The apparent shift is related to displacement and proportional to the variation of distance between emitter and receptor. It is positive when distance increases and negative when distance decreases. In astrophysics, the distance variation may also be replaced by space expansion and result with the same effect.

The above results can also be expressed as the difference between emitted and observed frequency and divided by the observed frequency, corresponding to a positive frequency shift, $\Delta f/f_R$, called a redshift when light is involved, and where we have

$$z = (f_E - f_R) / f_R \qquad (3.4),$$

which may also be expressed as

$$1 + z = f_E / f_R \qquad (3.5)$$

or the equivalent positive wavelength shift, $\Delta\lambda/\lambda_E$, or redshift

$$z = (\lambda_R - \lambda_E) / \lambda_E \qquad (3.6),$$

which may also be expressed as

$$1 + z = \lambda_R/\lambda_E \qquad (3.7),$$

where $z = 0.1$ is the redshift equivalent in our above example. Then, equation 3.5 yields $1.1\, f_R = f_E$ and equation 3.7 will similarly yield $1.1\, \lambda_E = \lambda_R$ in which the only difference between the emitter and the receiver is the stretch factor. Contrary to positive shifts, the convention requires that a frequency or wavelength shift, with a negative z value, corresponds to a blueshift.

3.1.3 What Happens When a Balloon Separates Source and Receptor?

Figure 3.3 represents a wave emitter and a receptor similar to figure 3.1 but with one important difference: a spherical balloon represented in 2-D has been added to the figure. Again, if the emitter and the receptor, located anywhere across the balloon, are not moving, waves are transmitted with constant amplitude, frequency, and wavelength. But if we assume that the sphere is expanding in a homogeneous medium and that densities inside and outside the balloon remain equal or vary with a negligible difference, then as the sphere expands, both the emitter and the receiver move away from each other, so the distance between them increases and the relative signal speed V_W decreases. Therefore, the relationships between speed, frequency, and wavelength are equivalent to our example on figure 3.2. The point is that the signal entering on the right side of the balloon is stretched while the balloon expands, and this is in proportion to the total expansion velocity or to the stretched distance when reaching the observer: it is positively Doppler shifted or redshifted for light or any EM wave. The inverse would apply if the balloon was shrinking instead of expanding: the signal would be negatively Doppler shifted or blueshifted, in the case of a light wave or any EM wave. This situation may be transposed to an earthly observer measuring the redshifts of galaxies and clusters located on the other side of our local expanding bubble if and only if we respect the fact that light is not traveling in a medium like air or ether but in a low-density field where space is expanding and stretching EM waves. But let us be clear: no new space is created within the cosmos.

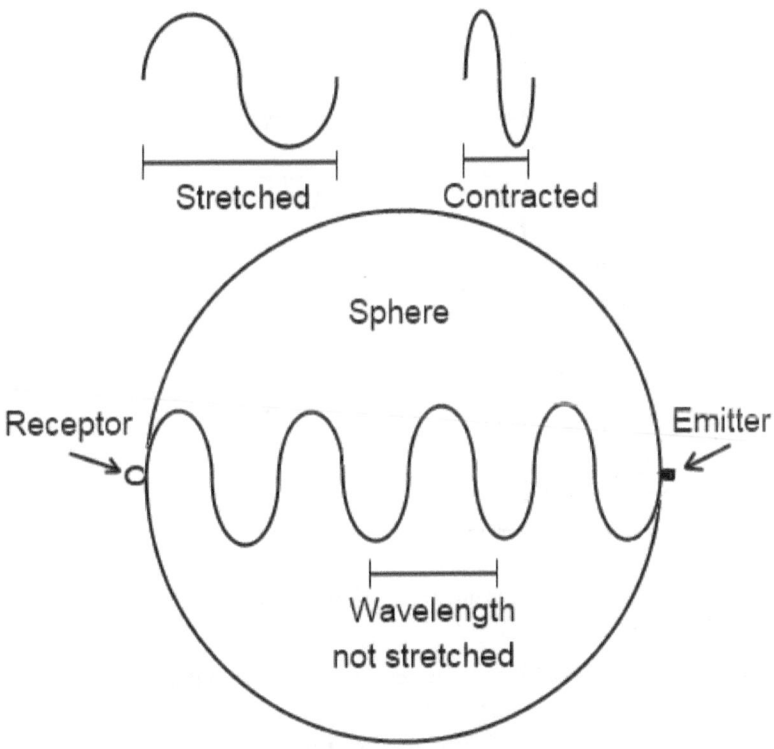

Figure 3.3 Expanding sphere between emitter and receptor

On the top left of the figure, a positive Doppler shift is illustrated as a stretched wavelength that occurs when the balloon expands, and on the top right, a negative Doppler shift, corresponding to a decreasing balloon, is shown as a contracted wavelength.

3.2   Cosmic Light Redshifts

An illustration of cosmic light redshift is presented on table 3.2 for a small portion of the electromagnetic spectrum, including the visible wavelength window between 380 nm and 740 nm and nearby part of the invisible ultraviolet shorter than 380 nm and infrared wavelength longer than 740 nm. A redshift of z = 0.2 is represented in the central column between reference absorption line values at z = 0 for the sun in column 2 from the left and redshifted absorption line values from faraway galaxies shown in the fourth column. The interpretation of redshift is questioned by some specialists, such as Halton Arp, who suggests the presence of intrinsic redshift not related to distance or expansion [2].

Wavelength nm	Sun	Z=0.2	Redshifted	
300	Ultraviolet		Ultraviolet	Invisible
400	Violet Blue		Violet Blue	380 nm
500	Green		Green	
600	Yellow Orange		Yellow Orange	Visible light
700	Red		Red	
800	Infrared		Infrared	740 nm Invisible

Table 3.2 Redshift illustrated from ultraviolet to infrared

We notice that redshifting increases with wavelength as we go down the table. For example, an ultraviolet absorption line at 300 nm in the sun is stretched by a length of 60 nm to 360 nm in our faraway galaxy, but an absorption line, at twice the 300 nm wavelength, in the yellow light of the sun at 600 nm would be stretched twice as much by a length of 120 nm to 720 nm in the faraway galaxy. Therefore, stretching or redshifting is proportional to wavelength [74].

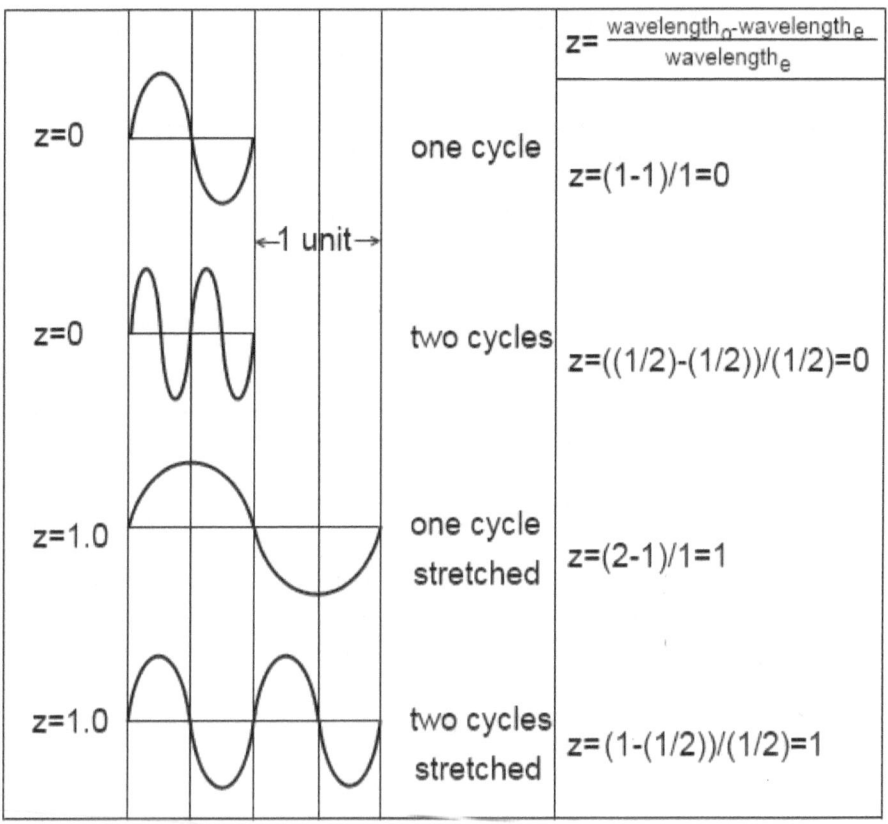

Figure 3.4 Comparison between one and two cycles for redshift z = 0 and z = 1

This redshift-related stretching is further illustrated in figure 3.4, where the values of z, indicated in the first column, are computed in the fourth column on the right using the wavelength equation 3.6, where subscript o means "observed" and subscript e means "emitted." One—and two-cycle waves are shown with the scale of one unit in the second column and where z = 0 at the top and z = 1 at the bottom, corresponding to a stretching that doubles the wavelength. The important point to notice is that the longer wavelength (lower frequency = one cycle) is stretched by one unit while the shorter wavelength (higher frequency = two cycles) is only stretched by ½ unit, confirming that redshift is proportional to wavelength.

Now, if we compare z = 0 with z = 1, we realize that energy is conserved since it takes twice the wavelength distance for z = 1, which corresponds to twice the time for the frequency. We know from equation 3.1 that E = hf, and so, for example, the frequency of one cycle per unit time at z = 0 will be cut by ½ and will take two time units after stretching at z = 1 and will have the same energy: E = h1 = h2/2. Therefore, light coming from distant galaxies and redshifted by space expansion conserves its energy since the same energy is distributed over a longer stretch corresponding, at constant speed c, to a longer period of time. Then, we have the following relationship that applies:

$$E = h(1 + z)f \tag{3.8}$$

or

$$E = h(1 + z)c/\lambda \tag{3.9}.$$

Thus, the expansion factor $(1 + z)$ redistributes the energy over the stretched EM wave, and so, in the cosmic microwave background, we should find all the emitted energy from all sources conserved in a stretched form.

3.3   Refraction across a Medium

Electromagnetic waves interact with the transparent media that they cross, and this interaction is called refraction and may be expressed as a refraction index n equal to the ratio of the speed c of EM waves in a vacuum to the speed v of EM waves in a medium:

$$n = c/v \tag{3.10}.$$

For example, the index of refraction in air is n = 299,792/299,704 = 1.000293, where c and v are in km/sec, and the index of refraction of zircon is n = 299,792/155,898 = 1.923.

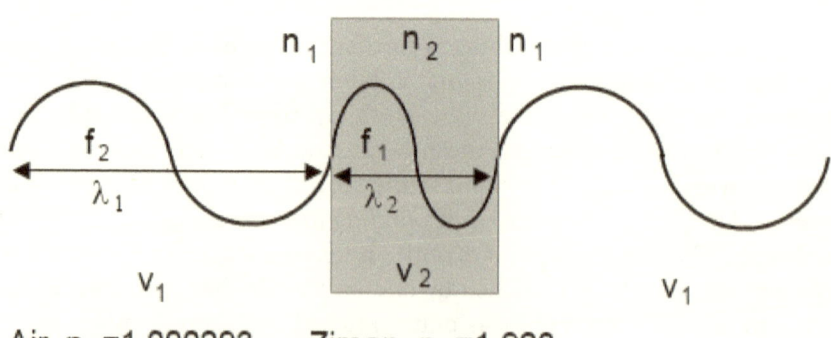

Air, n_1=1.000293   Zircon, n_2=1.923

Figure 3.5 Refraction across a denser medium

In figure 3.5, we apply these two refraction indices to an EM wave with an index of refraction n_1 in the air, entering a crystal of zircon with an index of refraction n_2 from the left and then reappearing in the air to the right. Since the index of refraction of zircon is almost twice as that in air and since the speed of EM waves in zircon is almost half of that in air, then we find that the wavelength in air is almost double of that in zircon and establish the following relationships:

$$v_1/v_2 = \lambda_1/\lambda_2 = n_2/n_1 \qquad (3.11),$$

where the frequency in the two media is the same $f_1 = f_2 = f$. That means velocity and wavelength in a medium are proportional to the index of refraction in the second medium. Thus, we expect to find a variety of refraction phenomena across higher-density clouds of gas and dust and filamentary structures in flat walls or in 3-D galactic and intergalactic space.

One interesting implication from the slower speed of EM waves in most media as compared to c is that neutrinos and neutrons can travel faster than EM waves in dense media since they don't interact like EM waves. It is theoretically possible to find neutrinos very slightly faster than the value of c if the latter has been underestimated in our solar neighborhood, especially since the speed of light, c, has never been measured in the extremely low-density bubbles of intergalactic space.

3.4   Refraction in a Prism

We have seen in equation 3.11 that wavelength is inversely proportional to the index of refraction. With figure 3.6, we see that the index of refraction n at the two interfaces of a prism varies inversely with wavelength λ: the white light is spread in its different wavelengths at the first interface, and this effect is amplified by the second interface from which we can see that the red component with the longest wavelength has the lowest refraction index since it is the least deflected and takes the shortest path; on the other hand, the blue component has the shortest wavelength and the highest refraction index since it has the greatest deflection and takes the longest path [25].

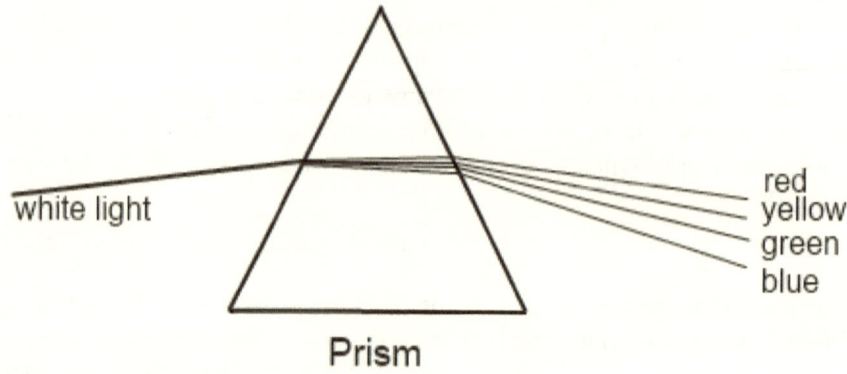

Figure 3.6 Refraction of light across a prism

For example, a glass prism may have a refraction index n = 1.535 for the blue wavelength at 380 nm and a refraction index n = 1.510 for the red wavelength at 740 nm, which makes a difference $\Delta n = 0.025$ in the index of refraction and explains the color separation.

Although the expected index of refraction across wedges of clouds and filamentary structures in galactic and intergalactic space is by far smaller than in the example above, it is not null and may affect significantly the mixture of light and other EM waves observed anywhere in the cosmos. This implies that light coming from high z regions and traveling over vast distances across a large number of bubbles may be doped by different wavelengths from a variety of its neighboring EM sources. It also means that this effect may be misinterpreted as gravitational lensing.

4. EXPANSION AND REDSHIFT, LIMITED CONTRACTION AND BLUESHIFT

After a summary of the bang time sequence for our bubble and for the 2dFGRS data slice across part of the cosmos, we will concentrate on bubble expansion and related redshift, followed by light trajectories and multiple bubble expansion and redshift. The link between redshift and distance will also be presented.

4.1 Observation of Data Slices across the Cosmos

In the past ten years, a substantial amount of redshift-space data has been gradually published to cover large volumes and slice volumes of the nearby cosmos [44]. The two most representative include the two-degree Field Galaxy Redshift Survey, which is second in size, referred to as the 2dFGRS and presented in figure 4.1, and the most voluminous and more recent ongoing Sloan Digital Sky Survey (SDSS) that will register data for over 100 million galactic objects.

The 2dFGRS was completed in 2003 and measured the photometry of about 380,000 objects and, among these, the spectrum of about 250,000 galaxies from two separate right ascension bands of 75°, with respective north and south declination of 7.5° and 15° in the direction of the galactic poles. The four-meter Anglo-Australian telescope was used for the survey in which two square degree pictures were recorded per exposure.

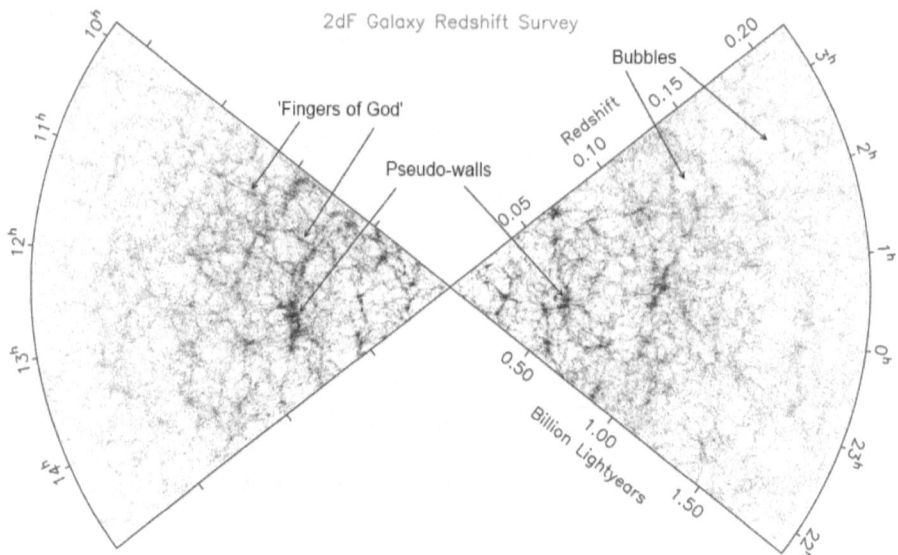

Figure 4.1 Two-degree Field Galaxy Redshift Survey

We are located at the center of this picture, where galactic north is to the left and where each tiny dot in the figure represents a galaxy. The scale is about 0.92 billion light-years for the redshift z = 0.1. We observe on this 2-D display of the 3-D slice a multitude of more or less flocculent bubble-like features of various size and shape, mainly circular or elliptical, which have been compared to soap bubbles, with galaxies floating all around the surface of the bubbles. The flocculent aspect is partly due to the piling up of 3-D data points on a flat surface. Pseudo-wall artifacts appear at about the same distance in north and south directions, and they occur where maxima of galactic data points and luminosity input are reached. Remember that we are not at a privileged location in the cosmos although large-density fluctuations are recorded. The quantity of information decreases toward the edge of these slices in proportion with distance and some spherical divergence. Many so-called fingers of God pointing in the direction of the observer are visible: these are generally associated with a line-of-sight spread of redshift values measured from clusters where galaxies are falling in toward the center of the cluster, as explained in figure 4.2.

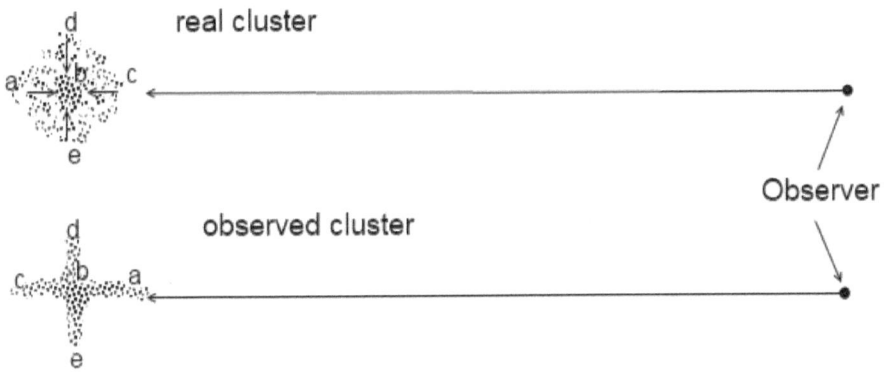

Figure 4.2 Real and observed cluster

The real cluster shown has one or more central massive galaxies surrounded by a large number of infalling galaxies at various velocities. Since the central galaxies near *b* have negligible velocities and that galaxies near *d* and *e*, perpendicular to the observer, are not Doppler shifted, then these galaxies along the line *d*, *b*, and *e* are observed in their real position. But for all the other galaxies, observation is skewed by light shifting: galaxies near *a* will be observed as blueshifted toward the observer across the central part, and inversely, the *c* galaxies will be redshifted away from the observer and on the other side of the central galaxies, thus forming the elongated shape in the observer's line of sight.

The 2dFGRS north and south displays are projections in a plane of data points getting thicker and thicker with distance since relatively small angles of 7.5° and 15° were used. A small angle slice across bubbles of various sizes and shapes and their projection on a plane is presented in figure 4.3, where redshift values are indicated on the top.

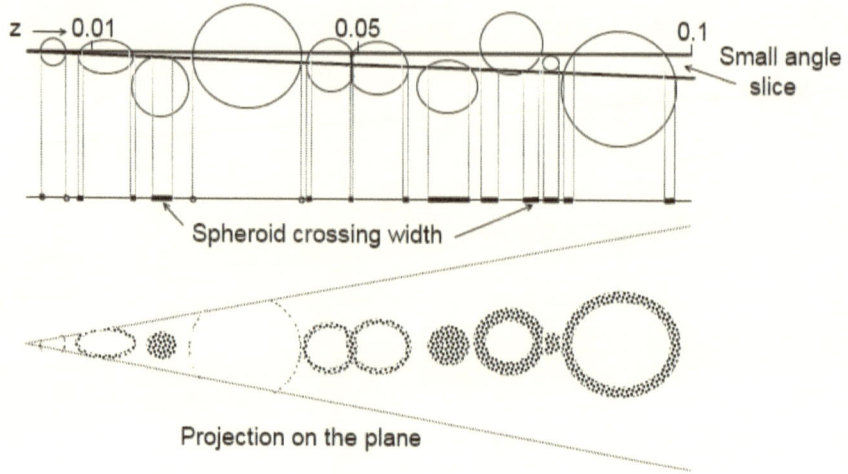

Figure 4.3 Small angle slice and projection on the plane

The width of the shell from the spheroid crossing the slice is projected on a line with black marks, which are in turn projected on a plane at the bottom. We note that the small angle includes more and more material as it becomes wider to the right, so the same-size bubble may look quite different, depending on location and on the portion crossing the slice. For example, the third bubble on the left is only grazing the slice, but it represents many data points.

What is the average part of the bubbles that small angle slices will cross? To answer this, consider figure 4.4 below.

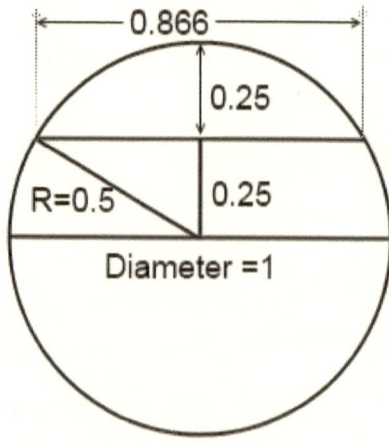

Figure 4.4 Average spheroidal bubble slice crossing

Since we are dealing with a thin slice crossing a large number of spheroidal bubbles, we realize that the average width of the slice is neither the maximum of 1 in the center or the minima of 0 at the top or bottom, but somewhere in between. We can see that the average width is halfway between the top or bottom and the center, where radius 0.5 is divided by 2 to yield a distance of 0.25 from the center: thus, from the Pythagorean theorem, the average crossing width is 0.866, or about 87%, for a diameter of one unit. This average crossing at 87% of bubble diameters implies an important correction to measurements of spheroidal or ellipsoidal bubble diameters from surveys such as the 2dFGRS. In other words, the surveys are underestimating the average radius of bubbles, and the corrected radius is $R_{corr} = R_o/0.866$, where R_o means "observed radius." This correction would not apply to SDSS data if the larger 3-D volume is used instead of small angle slices.

4.1.1   Expansion across Bubbles

Clusters are located around bubbles, as sketched on figure 4.5. As shown earlier, there may be as many as six primary hexahedral and eight secondary tetrahedral clusters in contact with each bubble for an average count of one hexahedral and two tetrahedral clusters per bubble.

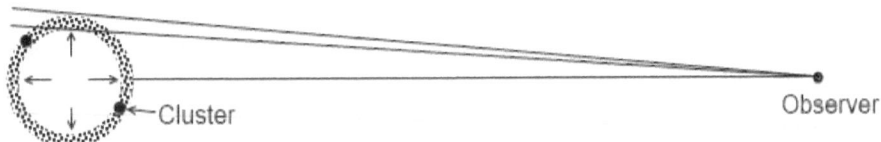

Figure 4.5 Expanding bubble and clusters

For example, if a bubble expands at a rate of 4,400 m/s across its diameter, then galaxies and clusters located on the expanding shell are moving away from one another, just like dots on an inflating balloon: the farther the dots, the faster they move away from each other. Note that this is a similar situation as the positive Doppler shift observed and defined with figure 3.3, where an expanding balloon was separating the emitter and the receptor. It is the space within the bubble that is expanding and producing the redshift, but no new space is created since the expansion is within the cosmos. So it may be qualified as local cosmological redshift directly associated with all the expanding bubbles that are interspersed with neutral nonexpanding bubbles. The emitted light coming from the left of the bubble is subject to the like of a partial track-stand as it moves and stretches across the expanding bubble. For an observer located far away to the right, this local expansion corresponds to an additional redshift Δz proportional to the expansion velocity v divided by c, the speed of light in vacuum:

$$\Delta z = v/c \tag{4.1}$$

So $\Delta z = 0.015$ between the far wall and the near wall of the shell. This light shifting across the bubble decreases gradually from the center to the top and bottom, and thus, the shape of the bubble is roughly preserved for the observer, although as seen in the line of sight, a large portion of the top and bottom may be dampened by extinction from dust or dense gas and crowding of galaxies. Therefore, in our model, each expanding bubble apparently stretches EM waves and contributes to the total z in proportion to its own expansion:

$$z = (v_1/c + v_2/c + \ldots) \tag{4.2}$$

where $v_1, v_2 \ldots$ are the expansion velocities of bubble 1, bubble 2, etc. This may be expressed as

$$z = \sum_{i=1}^{n} Vi/c \tag{4.3}$$

or as the general relationship:

$$z = v/c \tag{4.4}$$

which represents the sum of all the redshifts between light source and observer with v being the average velocity. The Hubble law, written as

$$H_o = v/d \tag{4.5}$$

in the big bang model, is modified in the multibang model to read as

$$H_o = v/(qd) \tag{4.6}$$

where q is a constant to correct distances for average space volumes where expansion is absent or replaced by contraction. Replacing the value of v from equation 4.4, we obtain

$$H_o = (cz)/(qd) \tag{4.7}$$

where our present-day Hubble constant is proportional to an average velocity over an adjusted distance.

Cosmic bubble parameters:			
Physical parameter	Number	Units	Remarks
number of bubbles	3.4924E+09	bubbles	for average size
average radius	28	megaparsec (Mpc)	1 Mpc = 1 million parsec
bubble density	0.02334	Mp/m³	proton mass/cubic meter
average radius	9.10E+07	l-y	light-years
average radius	8.6177E+23	m	
average volume	2.6808E+72	m³	cubic meters
ave. bubble mass	6.2570E+70	Mp	proton masses
proton mass	1.67E-27	kg	
ave. bubble mass	1.04E+44	kg	

Table 4.1 Cosmic bubble number and parameters

In column 1 to 4, we find the physical parameters, a number and associated units, plus some remarks. Since bubbles grow from invisible when they are nascent to an average of 28 Mpc radius at maturity, we present only average values in table 4.1. Bubble density is about ten times less than the cosmic average. The average mass of a single bubble is 1.04×10^{44} kilograms or about three times less than the mass of a single cluster. It includes the mass of the tenuous gas present at the center, where most of the volume is found, but it also includes gas, filamentary structures, and galaxies present in the fossil walls swept by the expanding bubble.

4.1.2   Maxibangs and Expansion

Observing all these bubbles from different surveys with their cohort of galaxies and clusters riding on their back, we are justified to ask, what happens if we contract these bubbles backward? To what origin in space and time is every individual bubble pointing? Or how old are they? We will see that they are all born at separate times somewhere in the cosmos and that the large spread of ages is associated with permanent and dynamic renewal cycles.

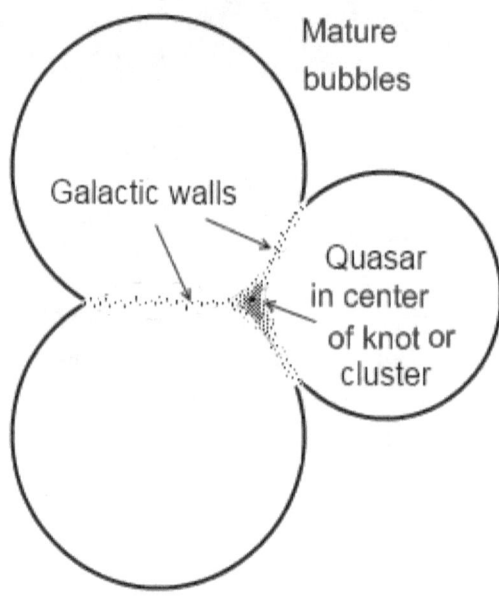

Figure 4.6 Mature bubbles with quasar in center

Little is known about the formation, life, and disappearance of quasars except that they are far away, often associated with powerful double and opposite jets of material, exhibiting a large spectrum that, depending on orientation, may cover the EM range from gamma rays to radio, and sometimes output energy comparable to hundreds of galaxies. We put forward the hypothesis that quasars are part of the final stage in the contraction of clusters of galaxies by gravity before collapse and rebound as a bubble.

A quasar forming at the intersection of four to six bubbles (only three drawn) is shown on figure 4.6. Mature bubbles have radiuses with an average of 31 Mpc (100 million l-y) and are separated by mostly flat walls composed of groups of galaxies in the process of being squeezed out and pulled toward the closest cluster. At the center of the cluster, a giant galaxy with an active galactic nucleus, or AGN, exhibiting a larger spectrum of EM frequencies is formed by accretion of smaller galaxies. The contraction process continues until a quasar is formed and exhibits its range of characteristics. After an average life of about 350 million years, quasars and their associates reach their final destiny in a gravity crunch from which they bounce back in a maxibang explosion as shown on figure 4.7.

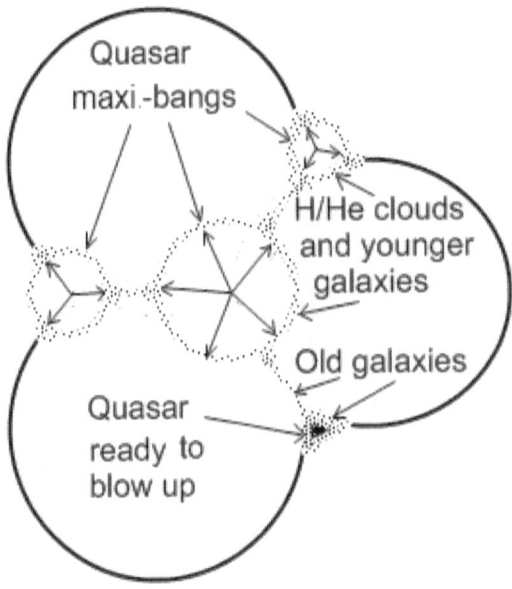

Figure 4.7 Maxibangs from quasars and expansion

Three quasar bangs are sketched in expansion, with shells of hydrogen and helium (H/He) gas forming the growing bubbles, which are gradually cooling down to temperatures where a new cycle of stars and galaxies will start to form. Older galaxies are also shown to indicate that older wall material will mix with the newborn and younger galaxies, so we may find local metal contamination in an otherwise pristine cradle. Also, a quasar surrounded with old merging galaxies is shown ready to blow up. But as the bubbles expand, they fill gradually a larger volume and will merge with their closest neighbors as depicted on figure 4.8. Streaming motion will be observed between neighboring bubbles.

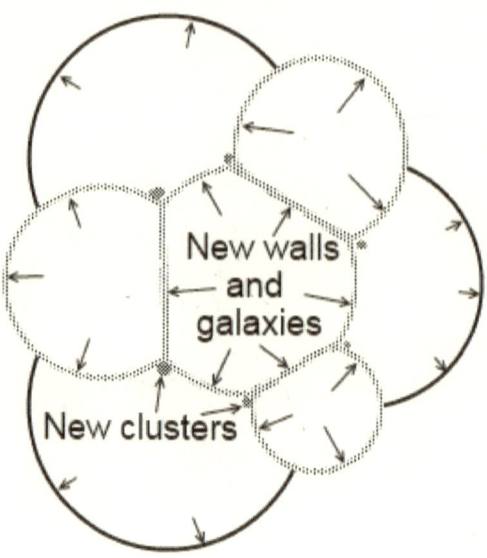

Figure 4.8 New walls, galaxies, and clusters

When two bubbles meet, they form a common dynamic wall that is either flat or curved inward of the larger bubble. Pressure will squeeze the newly twirled galaxies and star clusters toward the edges of the walls to form the next generation of galactic clusters [31]. Note that new bubbles are growing into the volume of older bubbles that are gradually destroyed although they may still be expanding at a lower pace.

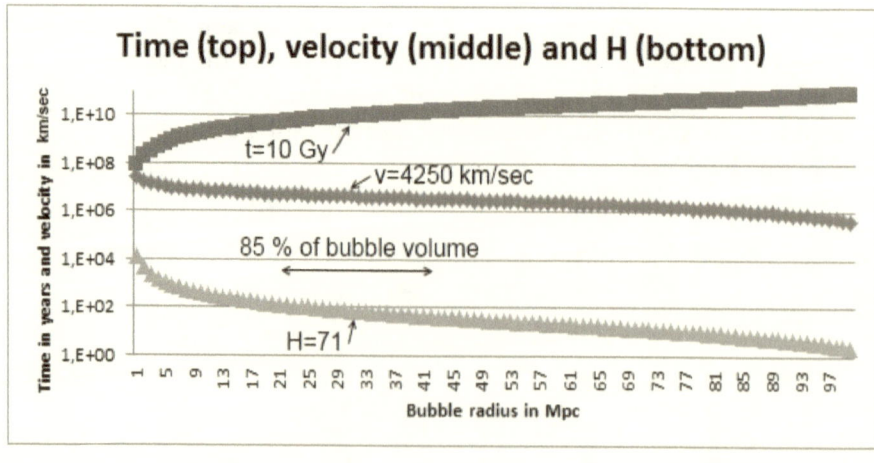

Figure 4.9 Time, velocity, and *H* values for bubble expansion

Bubble expansion is expressed in terms of age, velocity, and Hubble rate H on figure 4.9. The horizontal scale represents bubble radius from 1 to 100 Mpc although bubbles or residual shells with a radius larger than about 62 Mpc would be the exception and thus negligible in number. All radius values have been corrected using h = 0.71, where h = H_0/100 and where H_0 means "average observed Hubble constant" and is equal to 71 km/sec/Mpc from most recent surveys. About 85% of bubble volume is comprised between bubble radiuses of 21 and 42 Mpc, corresponding to bubble ages between 5.4 and 15.4 billion years with a bubble expansion velocity range from 3,400 to 5,230 km/sec and a Hubble expansion value H from 40 to 125. The curves on the figure are computed from a normalized escape velocity corrected for the kinetic energy losses from average bubble density material that is encountered and slowing down the expanding mass.

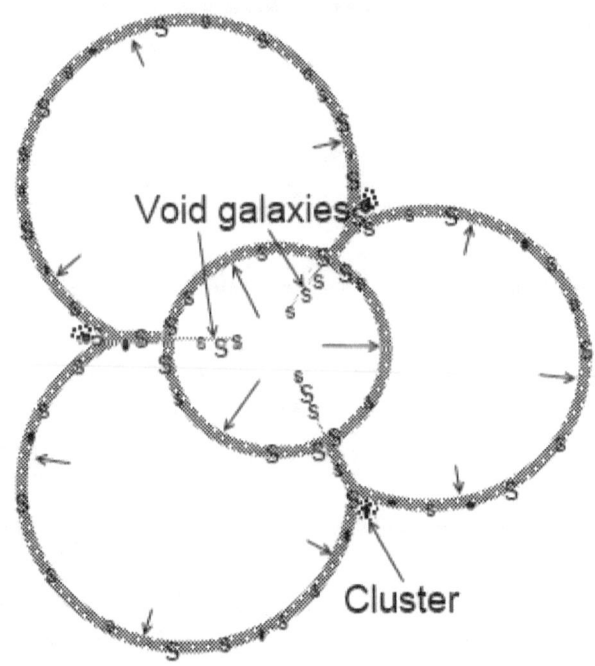

Figure 4.10 Void galaxies in expanding bubble

Larger clusters of galaxies are mainly formed in hexahedrons at the confluence of twelve walls and eight triangular sections, and smaller clusters are formed at the confluence of the six walls of four bubbles by the pressure of expanding bubbles combined to gravitational attraction between galaxies. A three-wall schematized version is shown on figure 4.10, where the main feature is the central bubble expanding into four (only three shown in 2-D) larger and older bubbles. Gas, dust, and lower-density material from the six walls between the older bubbles are swept away by the younger and more dynamic expanding shell, but some of the larger galaxies may be preserved in the process to form what are called low-density medium or void galaxies [46]. The latter galaxies are accelerated away by the expanding shell and may retain a fair amount of new gas from the passing shell and become active star-forming galaxies with old cores. Remember that 10% of the cosmic mass is found in the bubbles, which represent 90% of the cosmic volume but only about 5% of the galaxies, mostly as field galaxies located within the internal bubble voids [49].

Fossil walls, filamentary structures, and galaxies within the bubbles:			
Physical parameter	Number	Units	Remarks
number of walls	2.10E+10	walls	6 per bubble
mass of walls	1.64E+53	kg	5% of galaxies
volume of walls	6.5537E+79	m^3	cubic meters
mass of walls	9.83E+79	Mp	proton masses
density of walls	1.50	Mp/m^3	proton mass/cubic meter
Residual extremely low density bubbles excluding fossil walls:			
residual mass	2.01E+53	kg	in lowest cosmic density
residual volume	9.3624E+81	m^3	cubic meters
residual mass	1.20E+80	Mp	proton masses
residual density	0.01283	Mp/m^3	proton mass/cubic meter

Table 4.2 Fossil walls and extremely low density within bubbles

Physical parameters and associated values of fossil walls and extremely low-density region within bubbles are shown in columns 1 to 3 of table 4.2, with comments in column 4. As explained for the cosmic geometry, we expect an average of six fossil walls per bubble, and within these walls, we should find about 5% of the galaxies on a cosmic scale. The volumetric density of these walls is about one hundred times larger than that found in the most desolate voids, encompassing more than 99% of cosmic bubble volume. The term *fossil wall* means any residual wall material, left over when a bubble has ceased to expand, that is incorporated by a new bubble.

4.1.3 Light Faster across Bubbles

Void galaxies are reproduced on figure 4.11 in a blowup from the preceding figure. The expanding shell of a bubble leaves behind a low-density vacuum on its track; about 10% of cosmic density will fill this vacuum on average [81].

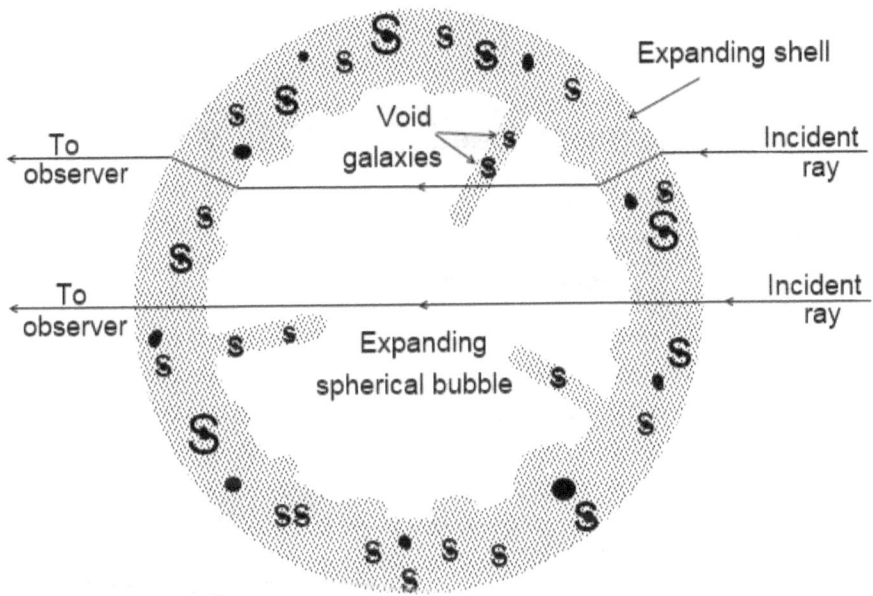

Figure 4.11 Light rays crossing an expanding shell

The speed of light, c, may vary slightly over the vast expanses of the cosmos. When an incident electromagnetic (EM) wave or light ray crosses the center of an expanding spherical bubble, then the maximum EM speed c is reached since it is crossing the lowest cosmic density and since it is located the farthest away from any gravity attractor or magnetic field. If we could measure c in such an environment, we would find the most accurate and fastest value of constant c. Incident rays crossing the expanding shell are refracted and thus slightly slowed down across the thickness of the shell, both at the entrance and at the exit, as shown on the upper part of the figure.

Light coming to us from faraway quasars will cross a large number of expanding bubbles where it is attenuated by absorption in the process. Most of the absorption in the Lyman-alpha forest of quasars is the result of EM waves crossing the shells in the central portion of the expanding bubbles, away from bubble edges where extinction from thick gas and dust may be present and completely obliterate the Lyman-alpha forest and other EM signals. In fact, Lyman-alpha forests of very far quasars will disappear because they are completely attenuated.

4.1.4   Light Crossing Time and Expansion

For a better understanding of light crossing an expanding bubble, we will look at the example illustrated on figure 4.12.

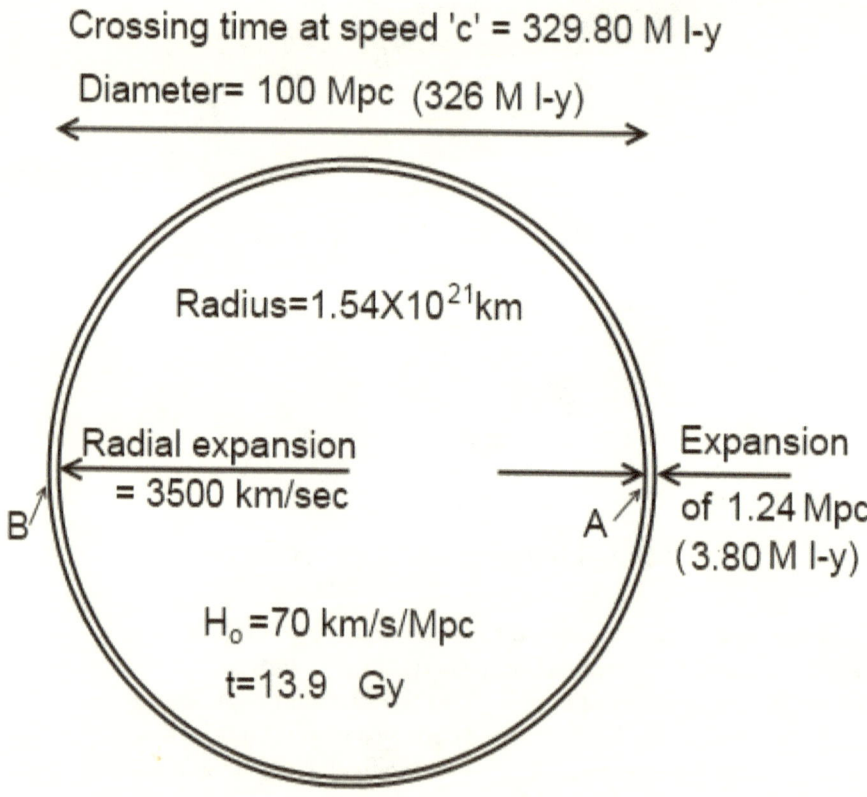

Figure 4.12 Light crossing time and expansion

The bubble has a large diameter of 100 Mpc (326 M l-y) with an expansion velocity of 7,000 km/sec corresponding to a redshift of z = 0.023, since z = v/c as per equation 4.4, and a space expansion rate H_o = 70 km/sec/Mpc. This is a radius equivalent to 1.54×10^{21} km with a radial expansion velocity of 3,500 km/sec.

Since light is traveling across the center at speed c of about 300,000 km/sec, it would take 326 M l-y to cross if the bubble was not expanding. But since the bubble is expanding radially at 3,500 km/sec, light will take 329.80 M l-y for its complete wall-to-wall trip, and as shown by the arrows, the bubble will increase its radius by an additional 1.24 Mpc (3.80 M l-y) during that time. From these values, we observe for the spherical model that the ratio of expansion to the original diameter $\Delta d/d = 0.023$ is equal to redshift z, so we have a new relationship:

$$z = \Delta d/d \qquad (4.8),$$

which may also be expressed as a ratio of the increase in radius over the original radius:

$$z = \Delta r/r \qquad (4.9).$$

Bubble diameter will increase more and more slowly in proportion to the Hubble rate of expansion H, and it will gradually take more time for light to cross a bubble, and z will increase proportionally.

These distance ratios remind us of the scale factor between the present and the past and related to z by the FLRW (Friedmann-Lemaître-Robertson-Walker) equation $1 + z = a_{now}/a_{then}$, where a is the scale factor, increasing monotonically as time passes. Thus, the mathematics of the FLRW metric would have to be adapted to our separate and specific expanding bubbles instead of the overall big bang cosmic expansion model.

The age of the bubble $t = 13.9$ billion years corresponds to the time taken to cross a radius of 50 Mpc at 3,500 km/sec. Only the first and most significant term is presented here for distances since the second term and followers would add a negligible amount of less than 1.2% to the expansion. It is important to note that because of expansion, light entering at the inner circle on the figure at A would exit at the outer circle at B. Also note that the above is just a numerical example since a bubble as large as 100 Mpc would most likely have an H value around 30 and an age of 20 billion years (Gy).

4.1.5 Average Light Trajectory across Cosmic Bubbles

For a generally spherical model of renewed bubbles across the cosmos one question is, where is light crossing a sphere on average? Figure 4.13 will give us a first answer.

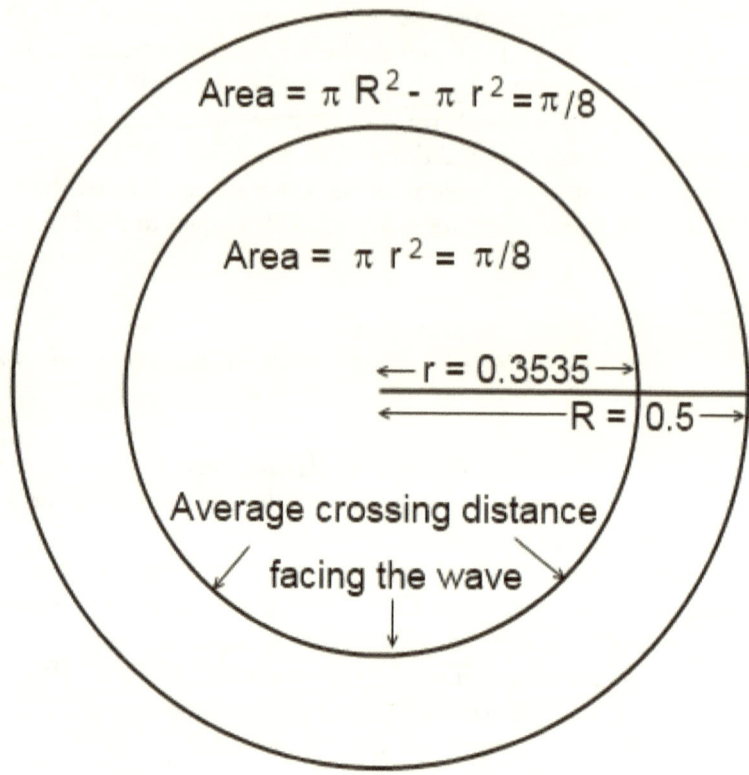

Figure 4.13 Average crossing distance facing the wave

Let us position ourselves as facing the EM waves coming at us across any spherical bubble. All that we can see from the sphere is equivalent to a flat circle in 2-D, as shown on the figure. From the geometry, we know that the average crossing distance will be located around a circle whose surface πr^2 is equal to the surface ($\pi R^2 - \pi r^2$) of the ring formed between that circle and the edge of the sphere. In our example, the larger radius, R, is equal to 0.5, and the inner radius is 0.3535, yielding both surfaces with areas of $\pi/8$ and a ratio of r/R equal to 0.7071, which is also the cosine of 45°. Therefore, the average crossing distance is at 0.7071 of the spherical radius, as also shown on the side view of figure 4.14.

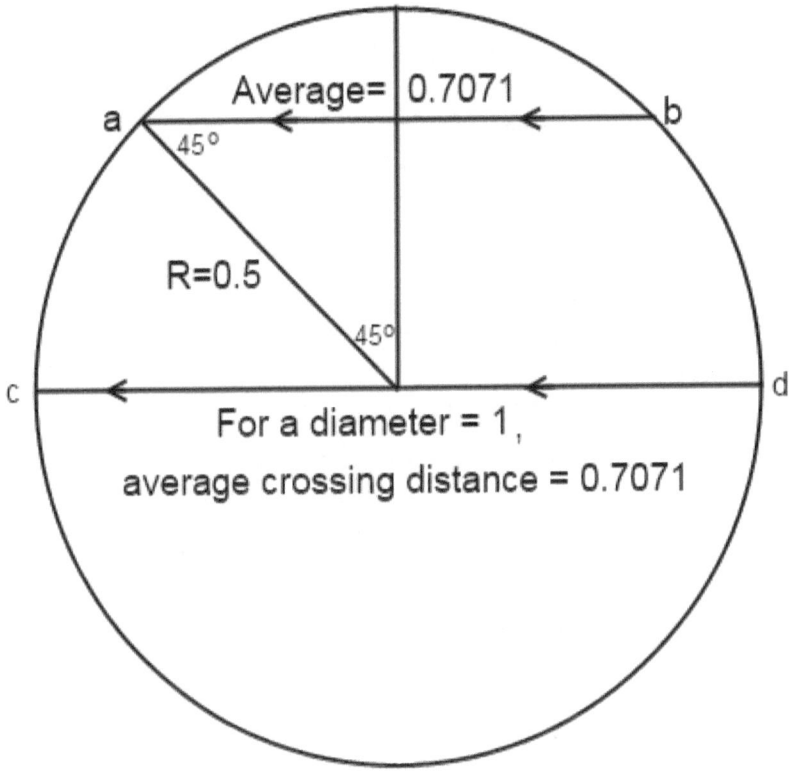

Figure 4.14 Light average crossing distance in a sphere

The side view shows a spherical model with a diameter of one unit between *c* and *d*, and also the 45° angle associated to the sine or cosine equal to 0.7071. So on average, light crossing a large number of bubbles will enter the sphere at point *b* and exit at point *a* and thus will only travel about 71% of the diametrical distance. If light was always crossing through the diameters of the expanding bubbles, then the rate of expansion would be 100% instead of the average of 71%, if we neglect bubble walls in a first approximation.

4.1.6   Local Bubble Expansion

With this average crossing distance in mind and remembering that our local bubble is obscured by the plane of our Milky Way galaxy, we will have a look at the expansion of our local bubble and its relationship with surrounding clusters [26] [66].

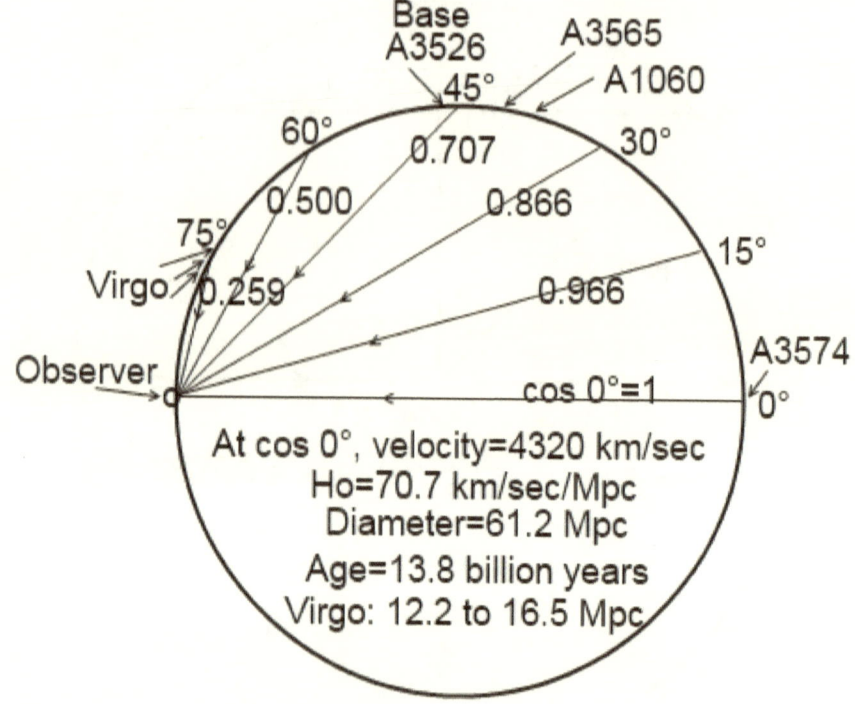

Figure 4.15 Local bubble expansion with base at cluster A3526

In the absence of a valid model for our local cosmic bubble (not to be confused with much-smaller-scale local bubbles in our galactic molecular clouds near the sun) and the 3-D distribution of galactic clusters surrounding it, we will use a spherical model where clusters are located at their average redshift distance on the surface of the sphere regardless of coordinates but respecting the angles from the observer. We consider light as crossing the sphere at an angle and exiting at the observer station on the left. Angles are viewed by the observer between the center and any point on the surface of the sphere, and they vary between 0° at the surface through the center and 90° at the surface near the observer. They are displayed in steps of 15° with the cosine of the corresponding angle, which gives the expansion ratio between 1 across the center and 0 at the observer. We find our average crossing ratio of 0.707 at 45° as expected, and we are using the Centaurus cluster Abell 3526 or A3526 as a base to locate other clusters.

Cluster A3565 from the Centaurus supercluster is found at about 40°, and Hydra cluster A1060 is located at about 35° from the observer but in a different direction. Also from the Centaurus supercluster, we find A3574 at about cos 0° the maximum distance of 61.2 Mpc on the sphere, assuming that all these clusters belong to the same bubble. Other clusters, like the Coma cluster and A1367, are located on a separate bubble adjacent to our local bubble and likely separated by a common wall in between [36].

For angles from 0° to 90°, every 15°, with cluster A3526 used as a base at 45° with z = 0.0102 and distance d = 43.3						
1	2	3	4	5	6	7
Angle θ in degrees	Cos θ expansion ratio	z = (cosθ*zbase)/ cosθbase	V = cz in km/sec	d in Mpc = (cosθ*dbase)/ cosθbase	Ho = V/d rate in km/sec/Mpc	t = d/V time in G years
0	1.0000	0.0144	4,328	61.24	70.7	13.8
15	0.9659	0.0139	4,180	59.15	70.7	13.8
30	0.8660	0.0125	3,748	53.03	70.7	13.8
45	0.7071	0.0102	3,060	43.30	70.7	13.8
60	0.5000	0.0072	2,164	30.62	70.7	13.8
75	0.2588	0.0037	1,120	15.85	70.7	13.8
74.3	0.2706	0.0039	1,171	16.57	70.7	13.8
77.2	0.2215	0.0032	959	13.57	70.7	13.8
78.4	0.011	0.0029	870	12.31	70.7	13.8

Table 4.3 Parameters used and computed for our local bubble

Parameters used to compute values for our local bubble on figure 4.15 are displayed in table 4.3. The first column gives the angles in degrees, with the last three values representing, from top to bottom, three different interpretations by de Vaucouleurs, Sandage, and Arp. The second column is the expansion ratio corresponding to the cosine of the angle and decreasing from the center to 90°, not to be confused with the Hubble rate of expansion H. The third column presents the average redshift z = 0.0102 at 45°, which is the basic observation by astrophysicists from which almost everything else is computed and extrapolated. The velocity V = cz in kilometres per second is tabulated in the fourth column and distance d in megaparsec in the fifth column with values decreasing from the center to 90°, just like velocities and distances between polka dots on an inflating balloon would increase. We find in the sixth column H = V/d, the Hubble rate of expansion in kilometres per second per megaparsec, which is a constant value within each spherical bubble but varies from bubble to bubble; for a larger bubble, we would find a smaller Hubble value, H < 71, and for a smaller bubble, we would find a large Hubble value, H > 71. Remember that *Ho* is an average based on a very large number of bubbles and it may largely deviate from the average in any bubble. Finally, in the seventh column, the age of our bubble, t = d/V, is shown as 13.8 billion years.

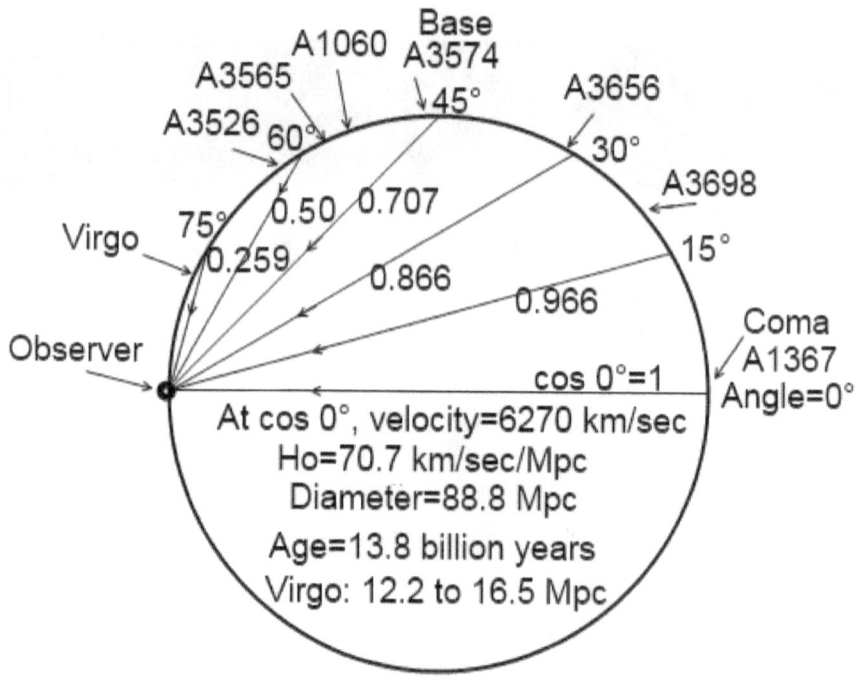

Figure 4.16 Local bubble expansion with base at cluster A3574

If we replace the base for parameter computations to cluster A3574 at 45° from cluster A3526 in table 4.3, we obtain the local bubble expansion model on figure 4.16. Since cluster A3574 from the Centaurus supercluster has a larger z value equal to 0.0148, corresponding to velocity V = 4,440 km/sec and a distance d = 62.8 Mpc, then the maximum at cos 0° is a velocity V = 6,270 km/sec and a distance d = 88.8 Mpc. All the other parameters, like *H*, age *t*, and the three Virgo distances, are unchanged. But the result is a local bubble that does not correspond to some published evaluations attributing a diameter of about 60-62 Mpc to our bubble. So if the larger model were real, then clusters A3698 at 20° and cluster A3656 at 30° from the Pavo-Indus supercluster could be projected on the sphere. Also, the Coma cluster and one of its larger neighbors, A1367, north of the Milky Way, would fall near the maximum diametrical distance of 88.8 Mpc, but between the observer and the Coma cluster, we find a large number of galactic groups forming a wall and suggesting the presence of a separate bubble near the Coma cluster.

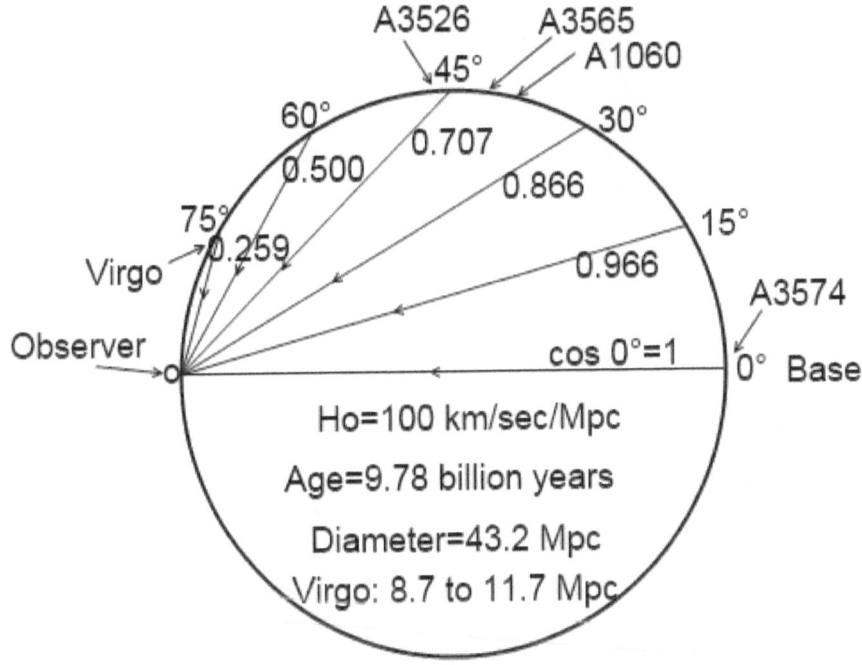

Figure 4.17 Local bubble expansion with base at cluster A3574

The local bubble expansion model shown on figure 4.17 has also a difference in the base for computations from that in table 4.3. This time, a position near cluster A3574 at 0° from the Centaurus supercluster is used, with a redshift z = 0.0144 corresponding to a velocity of 4,320 km/sec as on figure 4.15. This yields a diameter d = 43.2 Mpc and a Hubble rate H = 100 km/sec/Mpc equivalent to an expansion time t = 9.78 billion years. The Virgo distances are between 8.7 and 11.7 Mpc, which is much too low by most models based on standard candles like Cepheids and supernovae. This model must be rejected because distances are too small, age is too young, and the rate of expansion is too high.

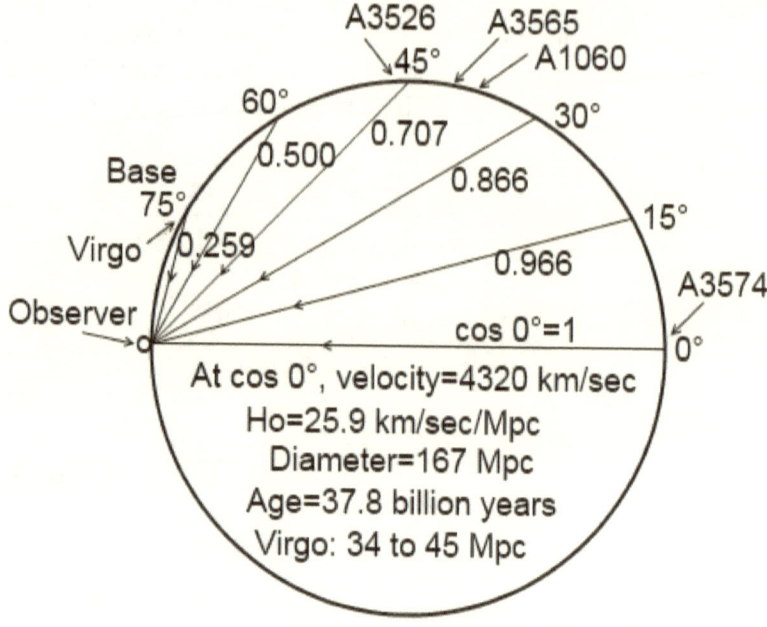

Figure 4.18 Local bubble expansion with base at Virgo cluster

The last comparison model for our local bubble is presented on figure 4.18, using a base at the Virgo cluster near 75° with an expansion ratio of 0.259. This base involves a redshift z = 0.00373 corresponding to a velocity V = 1,120 km/sec and, thus, a distance d = 43.24 Mpc. The Hubble rate of expansion is only H = 25.9 km/sec/Mpc in this conjecture, which yields a large diametrical distance d = 167 Mpc and an age of 37.8 billion years for the bubble. These results do not correspond to any known values relative to our local bubble: the age and diameter are way too large, and the Hubble rate is much too low. The Virgo cluster would be too far by more than a factor of 2. Therefore, this model is to be rejected and the first model on figure 4.15 seems to be the most reliable and closer to known calibrated values. The great advantage of this type of expanding bubble modeling is that it takes care of the very significant expansion ratio, in our own and neighboring bubbles, which is ignored in other paradigms.

4.2   Expansion and Contraction between Bubbles

After this presentation of our local universe made of an expanding void and surrounding clusters and galaxies, we will examine some aspects of the interaction between two and more bubbles expanding farther away than in the proximal cosmos. We will see how star clusters, galaxies, and galactic clusters are likely to form in a dynamic process that repeats itself again and again with each bubble expansion and clashing contact.

4.2.1 Dynamic Wall between Two Bubbles

The dynamics resulting from the contact between two bubbles is sketched on figure 4.19 with some of its attributes. When a bubble expands and reaches another expanding bubble, the expanding shells have left behind a void with a density about ten times lower than the average cosmic density [65]. This low-density medium is made up, in a proportion of about 12.5 hydrogen for 1 helium atom, of hydrogen and helium atoms drawn from the backside of the expanding shell, so a small density difference should be present between the front wave and the back of the moving shell, creating some density turbulence within the shell. This relatively small turbulence is amplified by the presence of up to twelve old vestigial walls and will produce eddies and twirls all around the shell, where cooling and locally increased densities will induce the formation of early stars and star clusters.

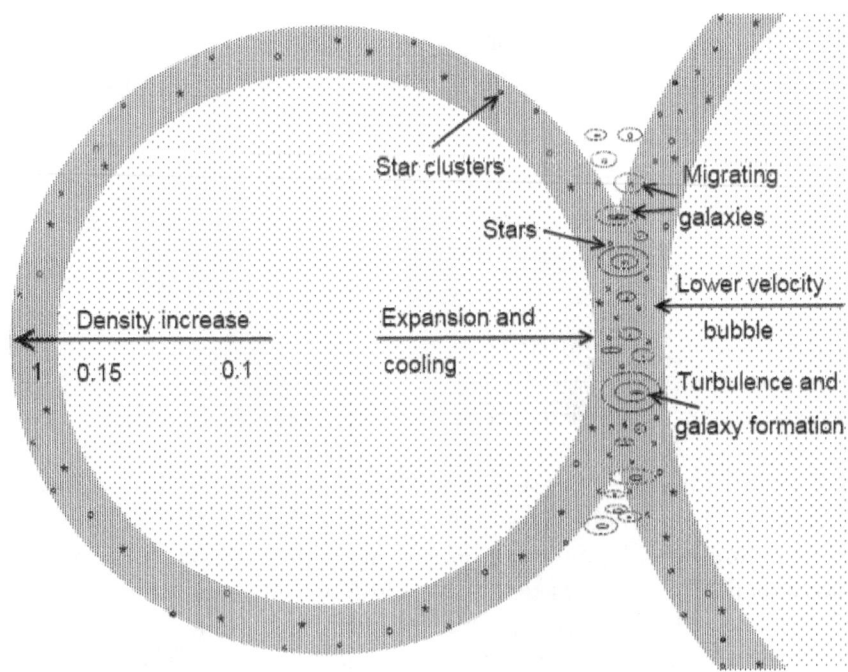

Figure 4.19 Turbulence and galaxy formation between two bubbles

As the two shells interpenetrate each other to form a common wall, their respective kinetic energy is very high, with velocities of a few thousand km/sec each, and this will produce high gas compression with large hurricane-like turbulence and eddies in a generally expanding and cooling environment so that the formation of galaxies will gradually proceed as the wall closes in with a flat or curved shape, depending on respective bubble size. Then, the residual kinetic energy left after the curling and whirling of galactic formation will push galaxies at lower velocities (streaming motions) out of the wall toward the concave triangular filamentary three-bubble junction from which it will gradually be squeezed out toward a concave tetrahedron at the junction of four bubbles to form a small galactic cluster that will in turn be squeezed out and pulled out to its final destination in a concave hexahedron at the junction of six bubbles where a large cluster made up of galaxies and hot gas will combine an equivalent critical mass of about seven thousand galaxies and start a new local cosmic cycle in a maxibang.

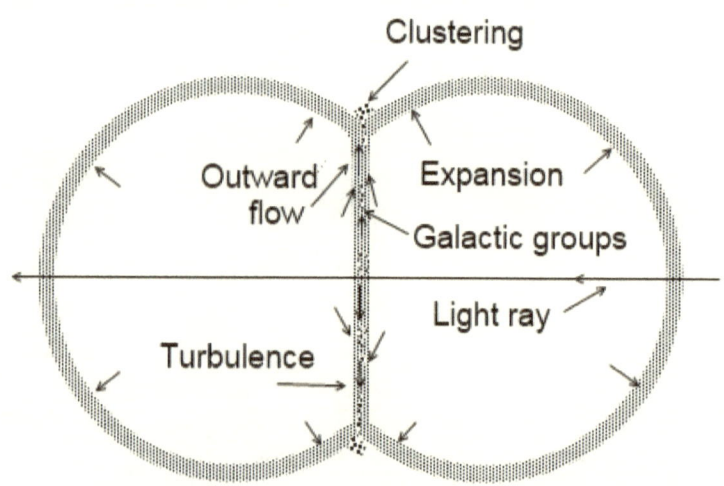

Figure 4.20 Wall formation between two expanding bubbles

Galactic formation between two expanding bubbles is presented on figure 4.20, where a light ray is shown crossing the two expanding volumes and the central wall. Turbulence is present along the wall for the formation of galaxy groups like our own galaxy with Andromeda and surrounding galaxies. Arrows indicate bubble expansion and the outflow of gas and galaxies toward the clusters. Note that the area near the wall may represent a sector where expansion is diverted sideways and more likely decreased, absent, or slightly inverted, thus inducing some errors in distance evaluations based on redshifts.

4.2.2 Contraction Space or Blueshift between Three Bubbles

As we progress from simple to more complex models, many questions arise, like what happens when a light ray crosses three bubbles and two curved walls? After evaluating that bubbles include about 90% of cosmic volume which is partly extending, this leaves about 10% of the space for clusters, walls, and filamentary structures. This means that about 10% of cosmic volume is not redshifted; on the contrary, this 10% is either slightly redshifted, neutral, or blueshifted. Therefore, a good cosmic accounting will include in its equations a systematic neutral or blueshift correction to obtain the equilibrium between expansion, neutrality, and contraction with adjusted shift z, corrected distances, and averaged Hubble rate of expansion.

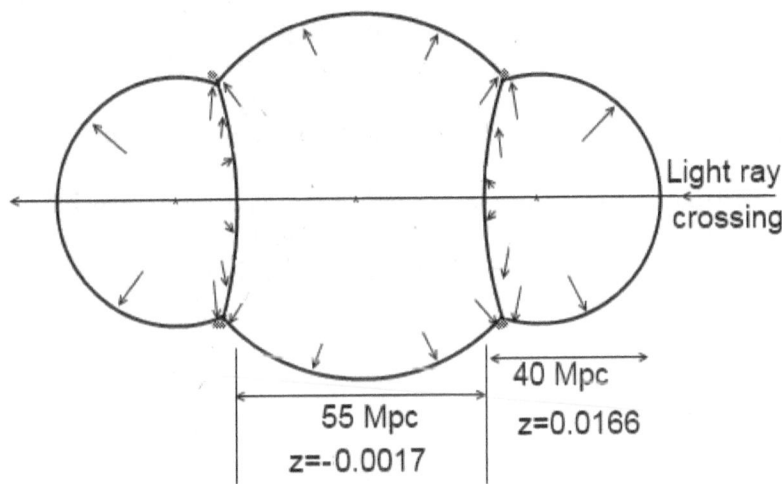

Figure 4.21 Expansion and contraction among three bubbles

Expansion and contraction among three expanding bubbles is schematized on figure 4.21, where a larger central residual bubble is flanked by two smaller and similar bubbles [30]. A dot indicates the original center of each bubble that is crossed from right to left by a ray of light. Arrows are roughly proportional to velocity vectors of expansion. As the two smaller bubbles collided with the larger one in the center, most of the kinetic energy was transformed to curls and twirls from which sprung stars and galaxies along the common walls. The smaller bubbles expand at a higher velocity and related kinetic energy so that the common wall is curving inside the larger bubble, as per equation 2.1.

Each small bubble has a width of 40 Mpc and a redshift value $z = 0.00933$ representing a positive expansion velocity $cz = 2,800$ km/sec and a standard

Hubble rate H = 70 km/sec/Mpc. But light crossing of the central bubble corresponds to a distance of 55 Mpc and a blueshift value z = -0.0017, which represents a negative expansion velocity cz = -510 km/sec (2 × -255 km/sec) and a low negative Hubble rate H = -9.3 km/sec/Mpc.

Now, we may compute the totals for a light ray crossing these three bubbles: the total distance is 135 Mpc with a net redshift value z = 0.01696 representing a total expansion velocity cz = 5,100 km/sec and a low Hubble rate H = 38 km/sec/Mpc. But if an astrophysicist uses a different model and takes the net observed redshift value z = 0.01696 to compute a total expansion velocity cz = 5,100 km/sec and divide by the average Hubble rate of expansion H = 70 km/sec/Mpc, then he would obtain a distance of about 73 Mpc instead of a total of 135 Mpc, which means that only 54% or about half of the total distance would be recognized, so the other half may be invisible and neglected in such a model based on the concept of a universal Hubble constant from a unique big bang event instead of multiple bang events averaging to the Hubble constant.

4.2.3 Tetrahedral Space between Four Bubbles

The tetrahedral space between four spheres and compacted bubbles is illustrated and compared in figure 4.22. The sketch on the left is similar to figure 2.7 but with the top sphere lifted to show the heart of the small cluster that would be nested in the center.

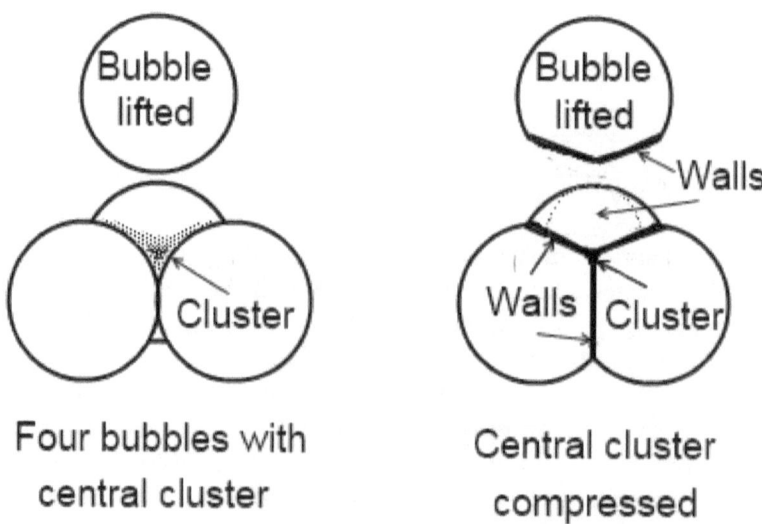

Figure 4.22 Tetrahedral space between four bubbles

On the right, the upper bubble is also lifted to show the reduced tetrahedral space in the center. The central cluster is compressed because the shells of the four bubbles have merged two by two for the formation of six walls and four concave triangular ducts that are guiding the spinning galaxies toward the more massive central cluster. The shape of these small tetrahedral clusters is illustrated in figure 4.23.

Figure 4.23 Cluster in acute tetrahedron

On the left of the figure, we present a view where all four apexes are well-defined and where the concave shape of the four sides is shown. A different view from the top of any apex is presented on the right side, with the concave curvature visible at the base. These tetrahedral structures generally correspond to smaller intermediate clusters that will eventually merge with the larger hexahedral clusters.

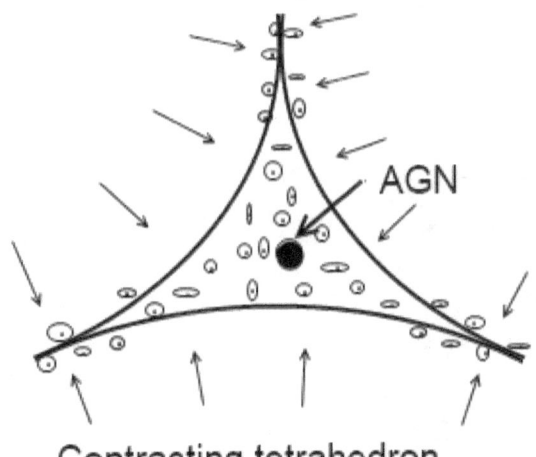

AGN

Contracting tetrahedron

Figure 4.24 Active galactic nucleus (AGN) in contracting tetrahedron

A 2-D sketch of a contracting tetrahedron is presented on figure 4.24, with a large galaxy near the center harboring an active galactic nucleus, or AGN, for short. We expect that most tetrahedral galactic clusters should have such an active nucleus when a number of galactic groups have started to accumulate in the center from the residual kinetic energy and pressure of the expanding bubble. Some of the infalling galaxies will be caught in the gravity field of the more massive center and may either collide and merge or gravitate for a period of time around the larger mass.

The virial theorem states that if equilibrium is present in a gravitational attraction system, such as a cluster of galaxies, then twice the average kinetic energy equals the average potential energy. But in our model, equilibrium for a virial radius may only be partly reached for galaxies gravitating close to the central mass of the cluster since other incoming galaxies are endowed with residual kinetic energy from bubble expansion. Thus, the high velocities of infalling galaxies are generally unrelated to the cluster's gravitational center and useless to evaluate the total mass of a cluster, and so dark matter is unjustified on this ground [57].

4.2.4 Hexahedral Space between Six Bubbles

We have seen in the Kepler conjecture that every sphere is surrounded by six hexahedral openings, and since each of the latter space is made up by the intersection of six spheres, we should find on average one hexahedron per sphere.

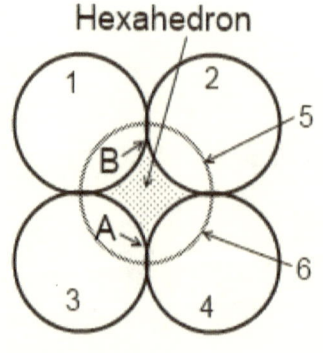

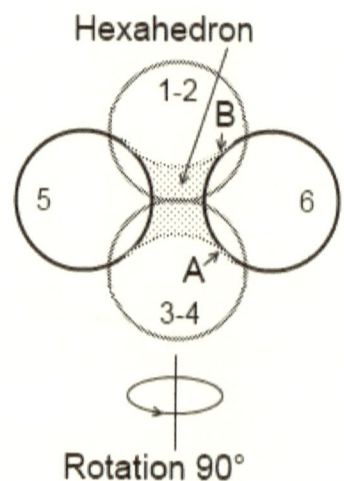

Figure 4.25 Hexahedron between six spheres

Figure 4.25 represents two different views from the hexahedron located between the spheres numbered 1 to 6. On the left, sphere number 5 is in the back, and number 6 is up front. The sketch on the right corresponds to a 90° counterclockwise rotation of the left drawing as seen from above: spheres 1 and 3 are respectively in front of spheres 2 and 4 after rotation. Letters *A* and *B* are used as markers to locate two different apexes before and after rotation.

Of course, the kinetic energy of bubble expansion will compress hexahedrons to a smaller size between six bubbles while galaxies are forming and migrating into these high-density clusters. Since the topology of all large-scale structures in the cosmos always corresponds to imperfect or irregular geometric systems, we expect to find spheres and bubbles of various sizes, so pentahedrons, for example, may also be found in observations instead of the expected hexahedrons.

Concave hexahedron

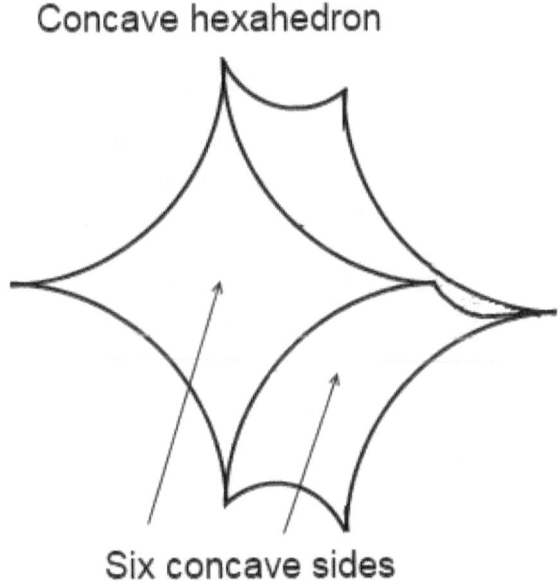

Six concave sides

Figure 4.26 Concave hexahedron produced by six spheres

A blowup of a concave hexahedron formed by six spheres is presented in figure 4.26. The concave curvature of the three sides facing us is shown. The shrinking from bubble expansion will vary according to the size of each bubble involved, and the hexahedron should eventually contract to a cluster where a quasar will form by merger of a large number of galaxies.

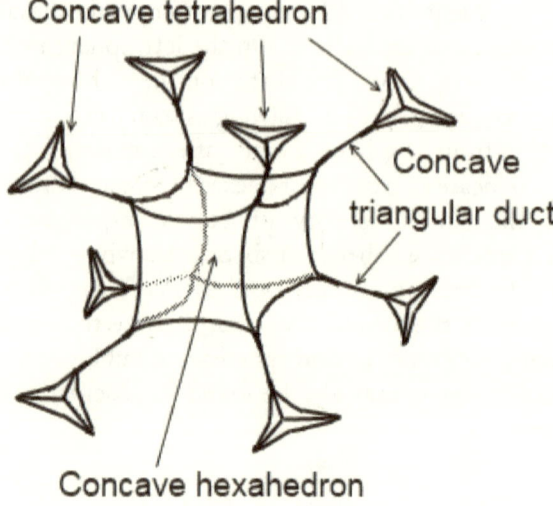

Figure 4.27 Concave hexahedron and feeders

In the face-centered close packing arrangement, a concave hexahedron may be fed by eight concave tetrahedrons via eight concave triangular ducts as shown on figure 4.27. We observe hierarchical structures leading to the hexahedron: groups of galaxies in walls between two bubbles followed by the triangular ducts and tetrahedrons. Note that two tetrahedrons located in opposite directions are replaced by two concave hexahedrons in the hexagonal close packing arrangement.

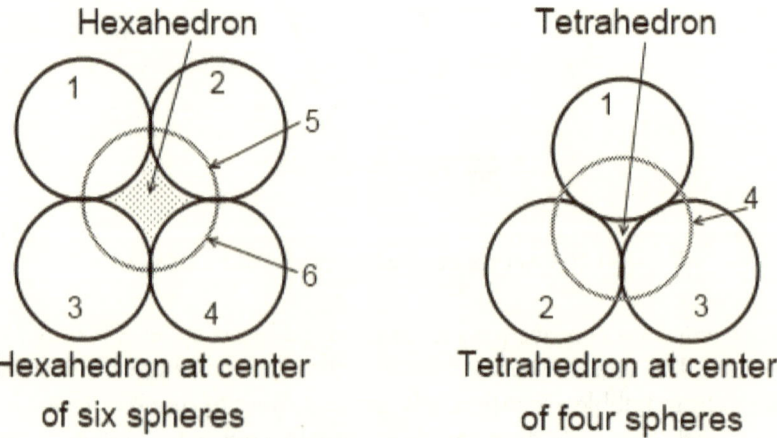

Figure 4.28 Comparison of hexahedron and tetrahedron size

A comparison at the same 2-D scale between hexahedron and tetrahedron is displayed in figure 4.28. We observe that the hexahedron occupies a volume many times larger than the tetrahedron. This explains why we expect tetrahedrons to form small migrating clusters and hexahedrons to contain the largest clusters with the development of large AGN and quasars that will eventually disappear under the effect of gravity compression as the most massive black holes and reappear as expanding new bubbles after a maxibang event.

4.2.5   Old Bubbles and Dodecahedron

But what happens to an old bubble that has expanded and clashed with as many other bubbles as possible? We have seen that each sphere in the Kepler conjecture is in contact with twelve other spheres, so we expect a mature bubble to have a maximum of twelve flat walls in common with its neighbors, thus forming a dodecahedron in the ideal geometric situation.

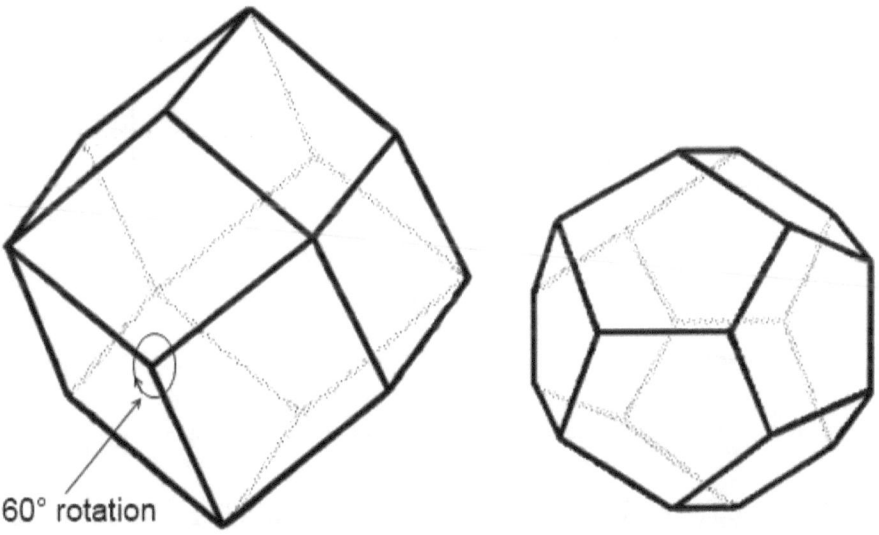

60° rotation

Figure 4.29 Dodecahedron or maximum number of bubble walls

Two different configurations are presented in figure 4.29 for a dodecahedron. On the left, we find a dodecahedron composed of twelve squares and corresponding to the face-centered close packing arrangement of the Kepler conjecture. A 60° rotation of three of the square faces would yield the hexagonal close packing arrangement. On the right, the twelve pentagons forming a dodecahedron are not corresponding to the expected local topology of a mature bubble.

4.3   Expansion and Contraction Intervals

A sketch of light rays crossing many bubbles is presented in 2-D on figure 4.30. Smaller and younger bubbles grow faster than older ones with consequent higher redshift rates and a shorter-lived youth. On the contrary, older bubbles grow more and more slowly with lower redshift rates and an extended life. Every new bubble after a bang takes up some space volume from up to twelve old bubbles. There is approximately one maxibang every 3.5 years in the cosmos, but the probability to see one is slim due to the slow speed of light over cosmic distances. So, we could perhaps witness one in 3,000 years in the readily accessible part of the cosmos surrounding us if we keep track of all presently known quasars during that time.

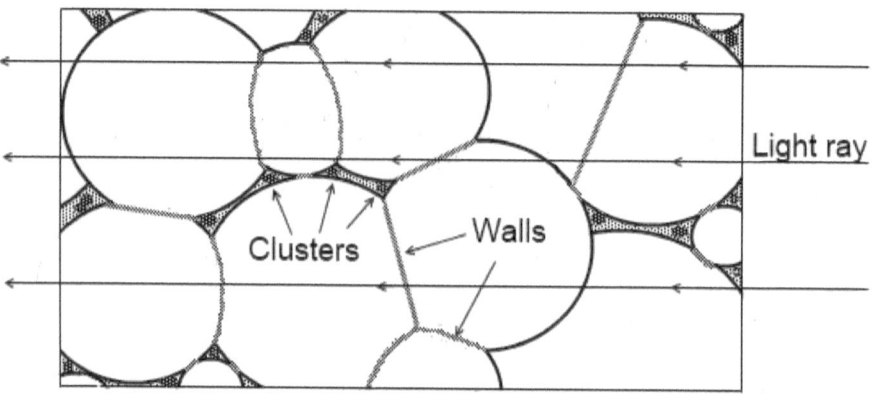

Figure 4.30 Sketch of light rays crossing many bubbles

As a permanent feature, small bubbles keep expanding with their high kinetic energy into older bubbles and also clashing with younger ones to form new galaxies where light will start on its cosmic journey. Light coming from faraway galaxies will generally cross groups of two or three bubbles in close contact because of a tendency for the repetitive presence of lattice layering of two or three layers across the cosmos, as previously shown in the Kepler conjecture for face-centered and hexagonal close packing arrangements.

4.4   Distance and Redshift

What could be a realistic constant for distance correction q in equation 4.6? The scale of distance shown in the 2dFGRS data slice of figure 4.1 was evaluated at 0.92 billion l-y for a redshift z = 0.1. But the more recent evaluation of distances in the SDSS indicates a value of 1.3 billion l-y for a redshift z = 0.1, which implies an increase of about 41% over the 2dFGRS estimate. This is not negligible and seems a good indication that distances are underevaluated in the big bang model. We saw in our example of figure 4.21 that, in a three-bubble model, an error in interpretation could yield a distance nearly ½ the real value in the model. In both cases, the range of correction or error is large and varies between 40 and 85%. So a first tentative evaluation of constant q in the equation of very large cosmological distances could be in the range 1.40 < q < 1.85 [54].

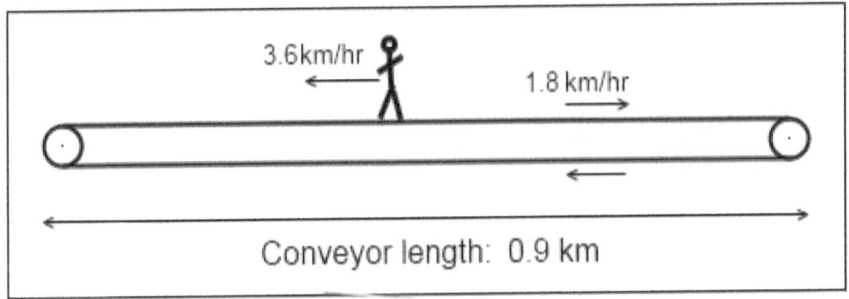

Figure 4.31 Jim walking in the opposite direction to the moving sidewalk

The following example on figure 4.31 should help to understand how the effects of space expansion, such as redshift and stretched wavelength, may be produced in a nonexpanding system that mimics expansion. Imagine a 0.9 km long moving sidewalk like those found in airports and Jim always walking at a constant speed of 3.6 km/hr or 1 m/s both on and off the conveyor belt. His constant speed of 1 m/s before reaching the conveyor represents steps of 1 m, each corresponding to the wavelength, and one step or cycle per second, representing the frequency. If the belt was not moving, Jim would cross its 900 m length in 900 seconds or ¼ of an hour. But the walkway is moving in the opposite direction at 1.8 km/hr or 0.5 m/s, so he needs twice the time, or ½ hour, instead of ¼ hour to cross 0.9 km. Thus, in ½ hour, he moves forward 1.8 km and backward 0.9 km for a total distance of 2.7 km, or 2,700 m, which is divided by ½ hour, or 1,800 seconds and yields a stretched wavelength or step of 1.5 m. When Jim's speed of 1 m/s is divided by the observed wavelength equivalent of 1.5 m, we find a frequency of 0.66 step or cycle per second. Then, if we divide the belt speed by Jim's speed, we find the equivalent of a redshift where z = 0.5. The moving belt has the same effect as expansion although nothing expands.

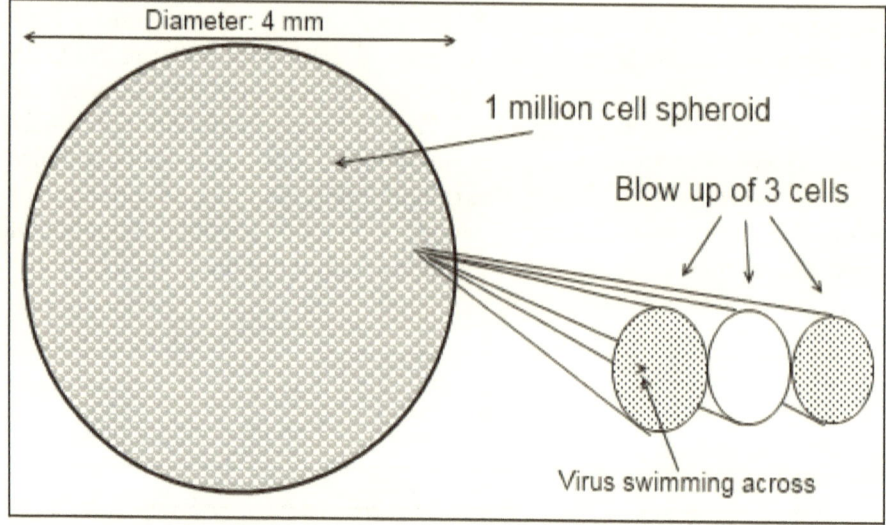

Figure 4.32 Self-regulation of growth cells in a 3-D spheroid

Laboratory studies on the growth of tumor cells in 3-D has demonstrated that a maximum diameter of about 4 mm and the formation of about 1 million cells may be reached for isolated spheroids [33]. Then, self-regulation takes over by controlling nutrient supplies to the far center, but new cells continue to replace older dead cells within the spheroid. This may be compared to our cosmic self-regulatory model in which older bubbles are replaced by newly born and expanding bubbles whenever gravity takes over and crushes a large galactic cluster to rebirth.

Now, imagine that half the cells in the spheroid are growing and that the other half is being recycled in such a way that equilibrium is maintained both in spheroid size and total number of cells, then a virus swimming at constant speed across the spheroid would take a longer time to cross all the growing or expanding cells, one out of two. If we assume that material is recycled along the walls between cells, then no contraction and no or very little blueshift equivalent would be associated with the other half of the cells.

Therefore, the expanding half of the cells would produce a redshift equivalent or a sum of small redshifts and would only give the measure for half the distance across the spheroid. This suggests doubling distance measurements, and so a larger value for constant q, closer to q = 2 is suggested. Note that our concept of bubble has some similitude to cells growing and dying in a spheroid with common walls and recycling. Of course, the cosmic concept has little to do with the exploding or popping action of soap or other gaseous bubbles. But we can conceive a cosmos in equilibrium where about half the space is filling up by expanding shells while the other half is made of nonexpanding dead or dying shells whose material is squeezed out along the walls into the clusters.

4.5   Cosmic Microwave Background

The cosmic microwave background or CMB [37] is the cemetery of all emitted photons that have travelled many times back and forth across the cosmos and stretched by an average z factor of about 1,100 over billions of years. It is the state of equilibrium between energy input and output reached by photons from which no additional heat is extracted. Among others, cosmic rays, high-energy particles from active galactic nuclei or quasars, and the Sunyaev-Zel'dovich effect play an important role in feeding back some energy to the CMB and preserve equilibrium [18]. It is the most important background radiation with a density of about 400 million photons per cubic meter [55]. The CMB is the only perfect blackbody since it includes all the electronuclear energy of the cosmos that is generated within the cosmic bright hole and recycled by gravity compaction.

The WMAP microwave anisotropy survey from NASA has mapped the CMB to the best coverage and precision up to the year 2010. Power spectrum maps showing a larger peak and smaller repeats of temperature fluctuations at the micro Kelvin scale and based on the angular scale of anisotropies have been published and interpreted in various ways. The main and secondary anomalies are likely related to the larger structures, including size and distribution of bubbles and clusters within the limits of resolution of the probe [82].

The CMB is definitely not the result of a unique big bang event followed by photon dissociation from electrons that would fill the present-day sky: this is physically impossible since such an event occurring about 370,000 years after the bang would form an expanding shell radiating away all that energy to a time distance of about 13.7 billion l-y (G l-y), and since light travels at 300,000 km/sec, then the Hubble expansion rate at 71 km/sec/Mpc (1 Mpc = 3.26 M l-y) would never be able to catch up with the faster-expanding shell of the runaway light, and with an open or flat model, there is no mechanism to return the light into the expanding cosmos. Also, there is no scientific or satisfactory basis for an inflation scenario except to justify a particular model. But luckily there is an alternative.

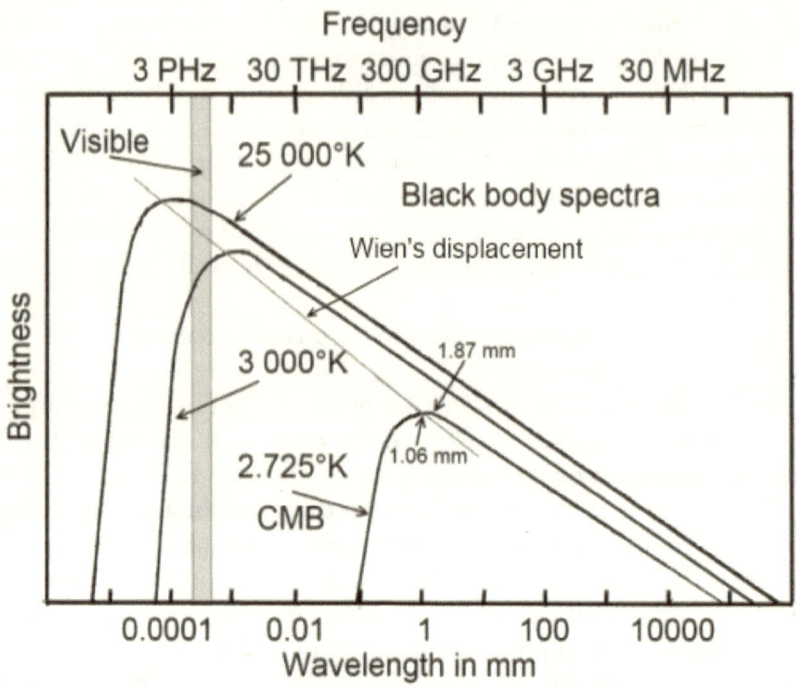

Figure 4.33 Blackbody spectra for three temperatures

Three different blackbody spectra are drawn on figure 4.33. The first is for a temperature of 25,000 K, representing a hot blue star at the top, followed in the center by 3,000 K, corresponding to the most common orange-red dwarf stars, and by 2.725 K from the cosmic microwave background. Our sun, with a surface temperature of about 5,800 K, would fit between the 3,000 and 25,000 K curves with a maximum in the visible range. A wavelength logarithmic scale is displayed at the bottom, and a frequency scale in Hertz is found on the top. The narrow gray band in the center corresponds to the visible part of the spectrum. It is flanked on the left by ultraviolet and shorter wavelength, and by infrared to radio wavelength on the right.

The height of the curves is proportional to brightness temperature, usually expressed as power per unit area, for example, in watt per square meter (W/m^2), which may also represent a density of energy or intensity function in the Planck radiation law applied to a blackbody spectrum. The CMB maximum for Planck's law is at a wavelength of 1.873 mm, corresponding to a frequency of 160.2 GHz. But the Wien displacement law states that the maximum wavelength in micrometers is equal to 2,897 divided by the temperature in K, which is $\lambda_{max} = 2,897/T_K$. So the CMB maximum, on Wien's slope shown as a gray line, is at a wavelength of 1.06 mm, which is a frequency of 283 GHz. We observe this displacement of the maximum blackbody emission from the three spectra on figure 4.33 that varies inversely with the blackbody's temperature. Thus, the maximum from a colder object corresponds to a longer wavelength.

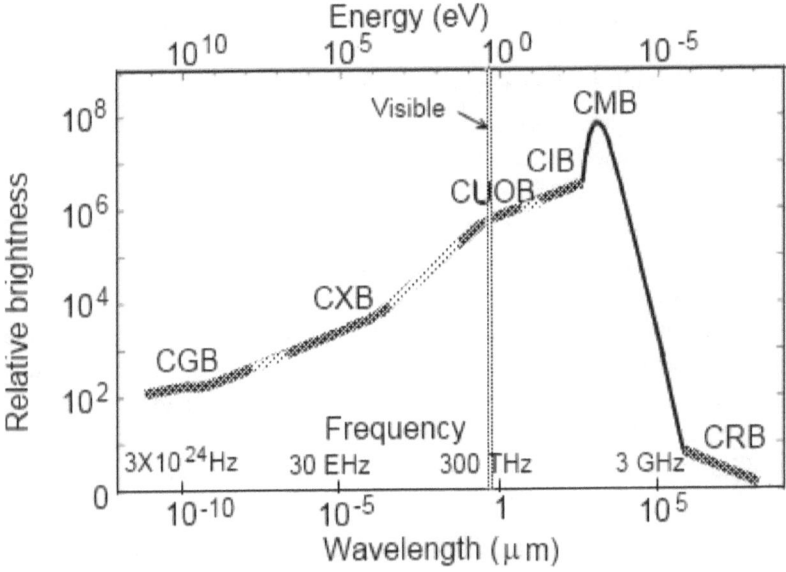

Figure 4.34 Extragalactic EM waves background

The blackbody curve, from the Cosmic Background Explorer (COBE) and Wilkinson Microwave Anisotropy Probe (WMAP) satellites with the millimeter peak of the CMB, is reproduced as a continuous black curve in figure 4.34, with the wavelength and frequency scales at the bottom and the energy scale in electron volt (eV) at the top [61]. The y-axis represents relative brightness with a log scale. This sketch is a summary of the more comprehensive distribution of all extragalactic EM waves that have been, so far, recorded by a variety of instruments sensitive to different wavelengths or photon energy. Visible light corresponds to the narrow vertical band.

White dotted bands along the spectrum are gaps in the actual data, and the gray dotted bands are a rough estimate of the recorded main EM wavelengths. The cosmic gamma ray background (CGB) is relatively continuous, but the cosmic x-ray background (CXB) could show two to three bumps and absorption gaps. The cosmic ultraviolet and optical background (CUOB) is more strongly attenuated by absorption at the high ultraviolet (UV) frequency end. A gap between optical and infrared needs further, more sensitive studies. Finally, the cosmic radio background corresponds to a low brightness, but little is known about the longer radio wavelengths. The common problem associated with all these background studies is the difficulty to remove or subtract light pollution from zodiacal sources in our solar neighborhood and from proximal galactic sources to obtain a clean background. A minimum threshold in signal-to-noise ratio must be reached before confirming the brightness of a specific background frequency.

This smoothed EM spectrum of the cosmos shows a general rise in the curve from high-energy to low-energy photons. Since the high-energy photons are more reactive, they travel shorter distances and are absorbed in gas clouds or space dust and are reemitted at lower frequencies in the infrared. The infrared wavelength photons may be stretched for hundreds of billions of years, as they generally have a longer mean-free path across the cosmos, until they reach the millimeter microwave background where they accumulate to a higher density, until equilibrium is reached with the amount destroyed and recycled by high-energy particles and by black holes sipping them up. The accumulation at the CMB peak is like a traffic jam produced by the repeated frequency blueshifting over billions of years of the microwave background photons by inverse-Compton scattering, as for example by the Sunyaev-Zel'dovich effect.

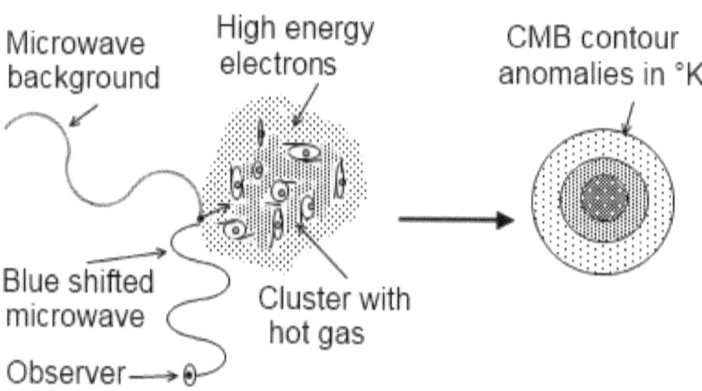

Figure 4.35 The Sunyaev-Zel'dovich effect across a cluster

The inverse-Compton scattering is illustrated in figure 4.35, where photons from the microwave background are blueshifted when they get boosted by high-energy electrons in clusters of galaxies containing hot gas and this in proportion with the density and thickness of the gas in the cluster [23]. The Sunyaev-Zel'dovich effect is a consequence of this shifting to higher frequencies of the microwave background photons, whereby a deficit in the number of photons is recorded for frequencies less than 218 GHz (wavelengths from about 15 to 1.38 mm) and an excess is shown for frequencies above 218 GHz (wavelengths from about 1.38 to 0.3 mm). So this small distortion on contour anomalies in micro K of the CMB will appear gradually fainter toward the center of a cluster at frequencies below 218 GHz and gradually brighter toward the center at frequencies above 218 GHZ. This effect is independent of redshift since it shows with equal intensity for similar-size clusters regardless of the z value of the clusters, but it is limited by instrument resolution for high z values. Consequently, the Planck satellite is searching for galaxy clusters up to a limited redshift of z = 1, with nine channels of frequencies both above and below 218 GHz.

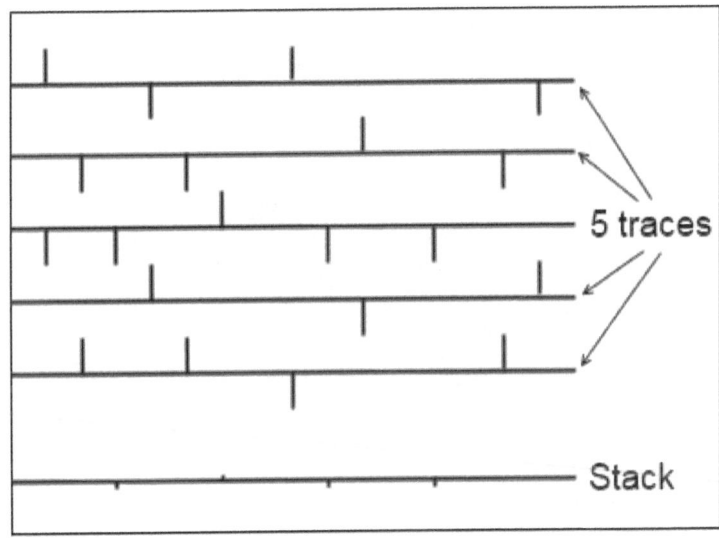

Figure 4.36 Stack of five signal traces

In figure 4.36, we present five signal traces and a stack, which is the sum of the traces divided by the total number. The process of stacking gives a surprisingly smooth result because data of opposite signs cancel out and noisy data spikes are attenuated by the number of signal traces. In nature, the blackbody spectrum of every star corresponds to a data trace with some absorption troughs that will add up to the signal from all the stars and emission peaks from gas clouds to yield after a long life of redshift the smooth cosmic microwave background where temperature variations are about one part in 100,000. These small variations are maintained by inverse-Compton scattering.

A small part of the CMB escapes by redshifting to lower EM frequencies in the radio range, and equilibrium is also maintained by the continuous black hole sipping of CMB photons. In fact, the CMB represents the stack of redshifted blackbodies equivalent to the output of 1 trillion times 1 trillion orange-red stars in the cosmos. It is glaringly easy to see the outcome: an extremely smooth redshifted blackbody background.

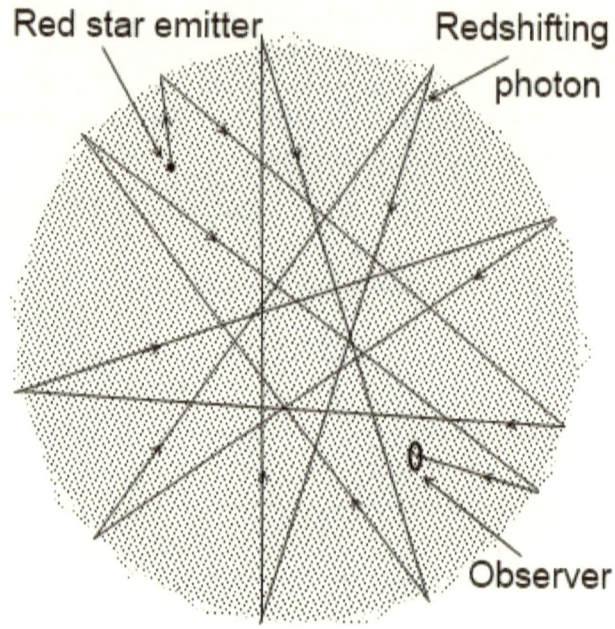

Figure 4.37 Trip of photons redshifted to CMB across the cosmos

A sketch is presented on figure 4.37 of the trip of a redshifted photon back and forth across the cosmos until the CMB is reached. A typical red-star emitter is shown with a part of the trajectory to the observer. Of course, it is the travel across expanding bubbles that produces the redshift or wavelength stretch. In fact, in the absence of any other interference, a 3,000 K average blackbody emitter will have its EM waves redshifted by a factor z = 1,100 to the CMB at 2.725 K. In order to accomplish such a feat, and since the average cosmic radius of 143 billion l-y corresponds to a redshift of about z = 10, then our photon must cross the cosmic diameter for an equivalent of about fifty-five times, which represents a total time distance of 15.73 trillion l-y.

The risk of crossing a hot gas cluster is not negligible over such a distance and may occur many times to boost back the microwave background frequency and maintain it at about 220 GHz, which is the average between Planck's and Wien's CMB maxima and also the transition frequency in the

Sunyaev-Zel'dovich effect. Since the microwave background is a near-perfect blackbody, it means that all frequencies across this 2.725 K blackbody are equally blueshifted by inverse-Compton scattering over very large distances.

If we use the above cosmic radius and the equivalent of 7 billion clusters, each with a radius of 13.4 million l-y and each harboring an average of 500 galaxies, then a photon will hit only 1 cluster in 31, and since we have about 1,907 clusters across a full cosmic diameter, then each photon should react eighty-seven times across the cosmos for a redshift value z = 20 or the equivalent of 4.36 hits for z = 1. Since we need an energy boost of 235 micro-eV to compensate a redshift of z = 1, then we should have an average increase of 54 micro-eV per photon scattering reaction and maintain CMB equilibrium.

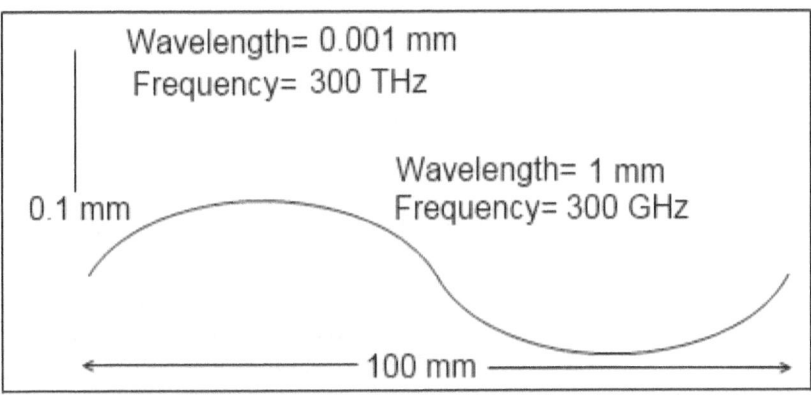

Figure 4.38 Comparison of 1 mm and 0.001 mm wavelength

What does a redshift around z = 1,000 look like in terms of wavelength stretching and frequency decrease? The answer is illustrated on figure 4.38, with a comparison of 0.001 mm and 1 mm wavelengths, shown at a scale one hundred times larger. On the top, a frequency of 300 THz, corresponding to the 0.001 mm wavelength, is drafted as very narrow vertical lines measuring 0.1 mm in width. At the bottom, a frequency of 300 GHz corresponds to a wavelength of 1 mm that is drawn as a 100 mm wave. The impressive wave stretching is proportional to its factor of 1,000 in frequency decrease.

If we apply equation 3.8, $E = h(1 + z)f$, where $h = 4.1357 \times 10^{-15}$ eVs, to the nonstretched 300 THz with z = 0 and to the stretched 300 GHz with z = 1,000 frequencies, we obtain in both cases E = 1.24 eV, which implies conservation of energy when bubble space expansion stretches EM waves across large distances within the cosmic spheroid.

5. STARS, GALAXIES, GROUPS, CLUSTERS, AND CLOUDS

Stars are the batteries generating the electromagnetic and nuclear energy in the cosmos while gravity is the power unit recharging them [22]. Galaxies are sweepers and concentrators of gas clouds that will generate the stars and, thereafter, crowd them up toward a central mass. The role of clusters consists mainly in building a nest of galaxies that gravity will eventually compress into a new bubble cycle.

Even in the most imaginable static situation, the dynamics are overwhelmingly present from all the forces and associated fields acting and counteracting simultaneously with EM waves—photons, neutrinos, and all the mass energy present. So a static mass energy state should be seen more like a snapshot of a dynamic process or the temporary dynamic solid state of a fluid or plasma.

Fluid dynamics is that part of mechanics that concerns the motion of liquids, gases, and viscous materials, including solids [45]. The behavior of a fluid is described by the Navier-Stokes equations, which are based on conservation of mass energy and of momentum, both angular and linear. A Reynold's number may be computed and assigned to a fluid to quantify the type of fluid flow: a laminar flow corresponds to a low number, and a turbulent chaotic flow with eddies will correspond to a large number.

5.1 Classification of Stars

Stars are divided in spectral classes designated by letters O, B, A, F, G, K, M, L, and T, in order of decreasing temperature, followed by numbers for each class from 0 to 9, which are increasing as the temperature decreases and described according to specific spectral characteristics that can be observed and measured from our earthly reference frame [77]. Table 5.1 is a summary of the spectral classes, with a range of temperature, mass, radius, and lifetime. At both ends, an extra class is added for marginal or extraordinary stars, like class W, which represents spectral type Wolf-Rayet and in which stars are part of the hot giant class O stars but specifically hydrogen depleted and so in the process of burning helium and metals (metals in astrophysics are atoms larger than helium). At the other end, we find class Y in the process of being defined by astrophysicists near the questionable limit between stars and planets that is found at about 1% of the mass of our sun or ten Jupiter masses and a star surface temperature that may be as cold as the human body, around 300 K, and thus definitely not a fusion combustion star.

Spectral class	Temperature in °K	Mass in sun masses	Radius in sun radiuses	Lifetime in million years	Relative fraction %	Mass X fraction%	Mass X percent X lifetime
W	50000	50	20	0,5	0,0000001	0,000005	0,0000025
O5	40000	32	18	1	0,00002	0,00064	0,00064
B0	28000	16	7,4	10	0,03	0,48	4,8
B5	15000	6,5	3,8	100	0,06	0,39	39
A0	10000	3,2	2,5	500	0,3	0,96	480
A5	8700	2,1	1,7	1 000	0,6	1,26	1260
F0	7500	1,4	1,4	2 000	1	1,4	2800
F5	6700	1,25	1,2	4 000	2	2,5	10000
G0	6000	1,04	1,1	10 000	3	3,12	31200
G5	5400	0,92	0,9	15 000	4	3,68	55200
K0	4800	0,8	0,8	20 000	6	4,8	96000
K5	4200	0,65	0,7	30 000	7	4,55	136500
M0	3500	0,45	0,6	60 000	10	4,5	270000
M5	2800	0,2	0,3	90 000	58	11,6	1044000
L0	2250	0,09	0,2	92 000	5	0,45	41400
L5	1700	0,075	0,15	90 000	2	0,15	13500
T0	1300	0,04	0,13	85 000	0,91	0,0364	3094
T5	900	0,025	0,11	80 000	0,1	0,0025	200
Y	400	0,01	0,1	75 000	0,001	0,00001	0,75

Table 5.1 Stars' spectral classes in terms of various parameters

In fact, the limit of hydrogen fusion is found near the middle of spectral class *L*, where a mass around 7.5% of the sun's is recorded and where fusion may be short lived, only a few million years since convection cooling and mass losses may hamper any further fusion. But with only a slightly larger mass above this limit, *L* or *M* stars just above 8% of the sun's mass could theoretically live for 1 trillion years or more from a slow core fusion and whole body convection process, although in the table, we limit the expected lifetime to around 90 billion years since it is unlikely that these stars would survive more than two or three cycles from the local gravity crunches at intervals of about 25 to 30 billion years without being recycled. And no L stars are known to harbor fusion with a mass less than 7% of the sun's.

Surface temperatures vary from more than 50,000 K for the *O* stars all the way down to less than 400 K for the new class *Y* stars. Some of the temperature limits and other characteristics vary in the literature as new data is coming in and interpretations are adjusted. Since the visible stars in the range of EM waves from 380 to 740 nm are the easiest to observe, they cover most of the table and subdivisions, but we should remember that the most numerous stars, at least in our galactic neighborhood, are from the fainter red dwarf *M* spectral class and have only 50% of the surface temperature, 33% of the mass, and 40% of the radius of the sun.

Star lifetime should vary from less than half a million years for the larger and hotter ones to possibly over 90 billion years for the slow hydrogen-burning red dwarfs, but the no-fusion brown dwarfs should behave more like large planets, with specific molecular characteristics and a variable lifetime dependent on their twin star or tidal neighbors. The relative fraction indicates that in our Milky Way galaxy, about 70% of all the stars are found in spectral class *M* with a surface temperature of around 3,000 K and with the longest lifetime of all stars. Since phenomena have a tendency to repeat themselves on a cosmic scale, we expect the preponderance of class *M* to be universal.

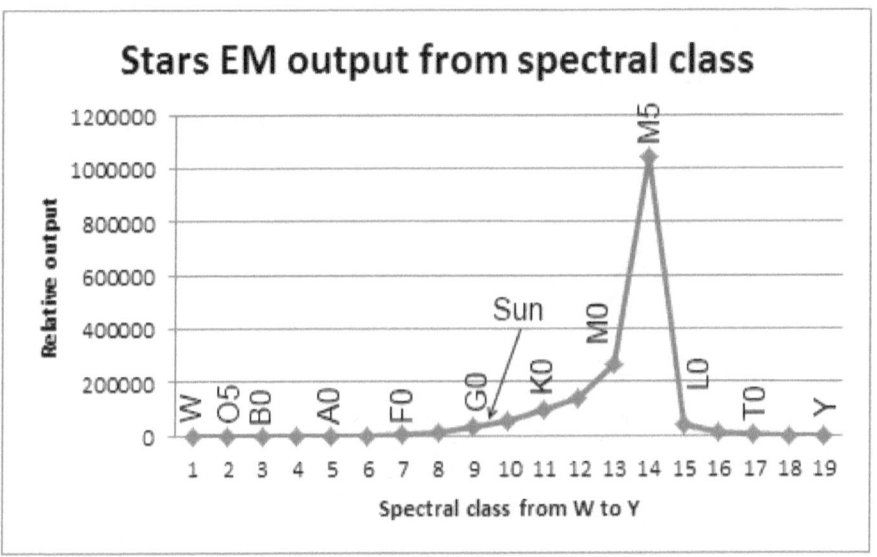

Figure 5.1 Stars' energy output in terms of spectral class

The seventh column of table 5.1 shows the product of mass with relative fraction, and since luminosity, given by the Stefan-Boltzmann law, is proportional to radius and temperature and, thus, a power function of mass M, usually between M^2 and M^4, then we find that spectral class M is the most important. The eighth column is represented on figure 5.1 and gives the product of column 7 with the lifetime that corresponds to cosmic star energy output in terms of spectral class. We notice an important asymmetric peak for class M in the distribution, but a Gaussian-bell-shaped graph would be more likely found if the study of stars included all EM waves equally and was not dazzled by visible light. It is the EM output from all stars and all spectral classes that are transformed and redshifted all the way to the cosmic microwave background.

5.1.1   Solar Radiation Spectrum

The blackbody spectrum of the sun with a temperature of 5,780 K is presented as an example of a star radiation spectrum on figure 5.2 with wavelength in nanometers on the horizontal scale and relative energy density on the vertical scale.

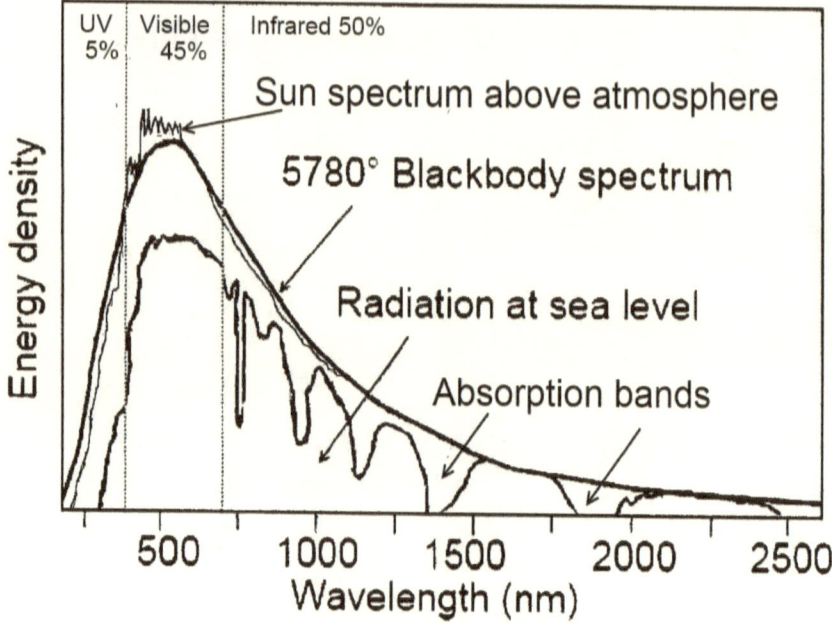

Figure 5.2 Sketch of solar radiation spectrum at sea level and above the atmosphere

A sketch of the typical and best known spectrum of a star as seen from one of its planets is shown. That is the sun as seen by earthlings. The sun spectrum above the atmosphere is very close to a blackbody in the infrared part of the spectrum but with an excess in the visible portion because of stronger emission and a deficit in the ultraviolet because of absorption [42].

On the contrary, as seen at sea level with the sun at the zenith, absorption is present in almost the whole spectrum except in part of the infrared. It is most important in the UV and visible but with very pronounced bands in the infrared. About 5% of UV electromagnetic waves are transmitted compared to 95% absorbed, whereas EM transmissions represent 45% in the visible and 50% in the infrared. In general, higher frequencies are more easily absorbed by atoms and molecules because they represent higher energies that react more easily according to the laws of quantum electrodynamics. UV radiation is absorbed by ozone (O_3), and other radiations are absorbed by water vapor, carbon dioxide, and the atmospheric gases.

In the next 10 billion years, the spectrum of the sun will gradually shift as nuclear fusion of hydrogen to helium increases to a maximum of 1.1 times its present value in about 1 billion years and then will shift to longer wavelength in the red over a 4-billion-year period as the sun swells to the size of a red giant while burning hydrogen at an average rate of 600 million tons per second in its upper layers above the core. So after depleting about 15% of its hydrogen fuel in the next 5 billion years, the sun will start to fuse its helium core to produce carbon and oxygen in a short period of 100 million years and expel its outer layers to form an earth-sized white dwarf star surrounded by its planetary nebula shell, expanding at an initial speed of less than 2,000 km/sec and gradually mixing with the surrounding clouds of star dust and molecular hydrogen to form new stars near the center of the galaxy, where the sun will have migrated down the spiral arm. In the next 5 billion years or so, the star will slowly cool and become a black dwarf almost invisible against the cosmic background, until about 10 to 15 billion years later, it is recycled with the galaxy in the crushing paws of the large next-door cluster and concentrated in the black hole of a newly born quasar.

5.1.2   Star Formation

If we want to understand the formation of galaxies and stars, we have to observe our own environment with all its curling eddies and whirling fluids and then transpose the model to the larger scales as the most likely scenario since similar processes are found at all scales.

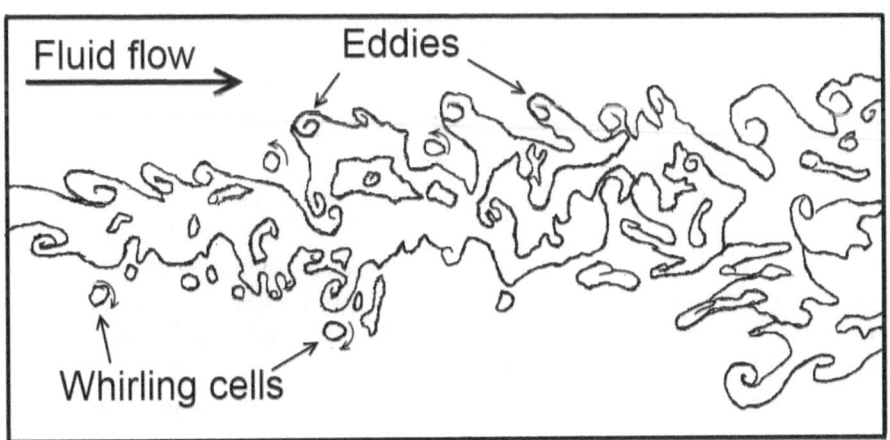

Figure 5.3 Turbulent fluid flow with eddies and whirling cells

The model drawn to represent a turbulent fluid flow on figure 5.3 could apply to any kind of fluid, gaseous or liquid. The flow is introduced from the center left of the sketch and moving near the center toward the right. We note an enlargement of the flow to the right as the energy is dispersed in the fluid, but the most fascinating feature is the formation of eddies in directions opposite to the flow and all around the flow in 3-D with the resulting appearance in 2-D of clockwise eddies at the bottom and counterclockwise ones at the top. Some whirling cells will form when an eddy escapes from the main stream, and the cells will keep the same direction of rotation.

Stars are known to form in galactic molecular clouds from the available material, containing a mass fraction of about 74% hydrogen, 24% helium, and 2% metals (all the other elements), corresponding to a mole fraction of 92% hydrogen, 7.5% helium, and 0.5% metals [63]. The fractions vary depending on the age and mixture of the local cloud within the more general galactic history. An older galaxy will concentrate a larger amount of dust and metals from its nucleosynthesis history and cannibal activities while a younger one will mostly proceed to fuse helium from hydrogen although some dust and metal contamination is likely always present from prior maxibang and bubble cycle.

Thus, stars in those molecular clouds should emerge when the rotating eddies, with a conical depression or cylindrical hole at the center, are separated from the larger cloud to become an autonomous rotating entity gradually losing kinetic energy until equilibrium is reached with gravity forces that will collapse the central mass and flatten out to a disk with its retinue of planets and satellites that have formed as smaller secondary eddies. Whenever two primary eddies are close by, they may gravitate around a common barycenter and collapse to form one of the 35% of binary stars in the galaxy.

Molecular clouds vary in size from 100,000 to about 6 million solar masses. Their temperature varies between 1 and 100 K with an average evaluated at 10 K, which corresponds to a low molecular energy that will facilitate the gathering work of gravity over the masses involved in the formation of stars and planets. The particle density varies from 100,000 to 100 million per cubic meter and will locally reach more than 1 billion, especially away from the high-energy radiation and within the protected center of a cloud. Molecular clouds are also called star nurseries when they have the right characteristics to generate stars that may be observed in the making. Groups of proximal stars may form as open clusters with weak gravitational links, so some members may escape.

5.1.3 Main Sequence to Neutron Stars and Black Holes

The main sequence is the continuous band of stars that stands out on color-magnitude plots known as Hertzsprung-Russell diagrams [68]. The sequence represents all these newly born stars, also called dwarf stars, producing energy from the nuclear fusion of hydrogen to helium in their dense core at a temperature of at least 10 million K. These stars keep an equilibrium between the outward thermal pressure generated by fusion and the inward compressive pressure from gravity for a short period of time of less than 1 million years for massive stars with efficient radiation-convection energy transport and an extremely long period that could theoretically exceed 1 trillion years for wholly convective subdwarf stars. Slowly rotating or nonrotating main sequence stars with less than 1.4 solar mass, the Chandrasekhar limit, will form helium from the proton-proton complex chain reaction, and above that mass and 17 million K, the carbon-nitrogen-oxygen, or CNO, cycle is more efficient at producing helium from hydrogen.

With the exception of the long-lived subdwarfs, which may avoid a cataclysmic end for a long period of time, and stars with a mass less than the Chandrasekhar limit of 1.4 times that of the sun, some white dwarfs with a large companion and nonrotating stars larger than 1.4 times that of the sun may experience at least one explosive event in their lives. These events are called novae or supernovae, depending on their mass and related history.

Novae are recruits from the 35% of binary stars in the Milky Way or the equivalent in other galaxies. Usually, the gas from a red giant is siphoned by tidal gravity by its white dwarf companion, which builds up an accretion disk until the star reaches the Chandrasekhar limit, where fusion reactions occur in an abrupt manner between the core and the surface with a sudden rise in the luminosity of the white dwarf by factors as high as 1 million, followed by a gradual decrease. Any unprotected fusion reaction not far from the surface of a star may obviously be the source of high-energy x-rays or gamma rays.

Solitary stars like the sun and up to 1.4 times the mass of the sun may evolve as red giants when they start burning hydrogen in the layers above the core and later fuse helium to carbon-oxygen in the core. They may expel their outer layers at high wind speeds without the explosive death of supernovae. The remnant is a planetary nebula with, at the center, a white dwarf with a carbon-oxygen core that will gradually radiate energy until it becomes a black dwarf.

Figure 5.4 Transparent shell from supernova SNR 0509

Supernovae form the more cataclysmic class of stars in the family. Figure 5.4 shows a transparent shell remnant from supernova SNR 0509, also shown on the front cover, taken by NASA's Hubble Space Telescope. A white dwarf or neutron star left over from the progenitor star is visible in the center, and an expansion velocity of 5,000 km/sec has been measured [70]. For the more energetic supernovae, the expansion velocity may reach 30,000 km/sec, which is $1/10$ the speed of light. The amount of energy involved in the implosion and flash fusion of a few seconds represents from 1×10^{44} joules for the lower masses and up to many times that quantity for giant stars. The luminosity increase may be many billion times that of the sun, making the supernova brighter than a whole galaxy [78].

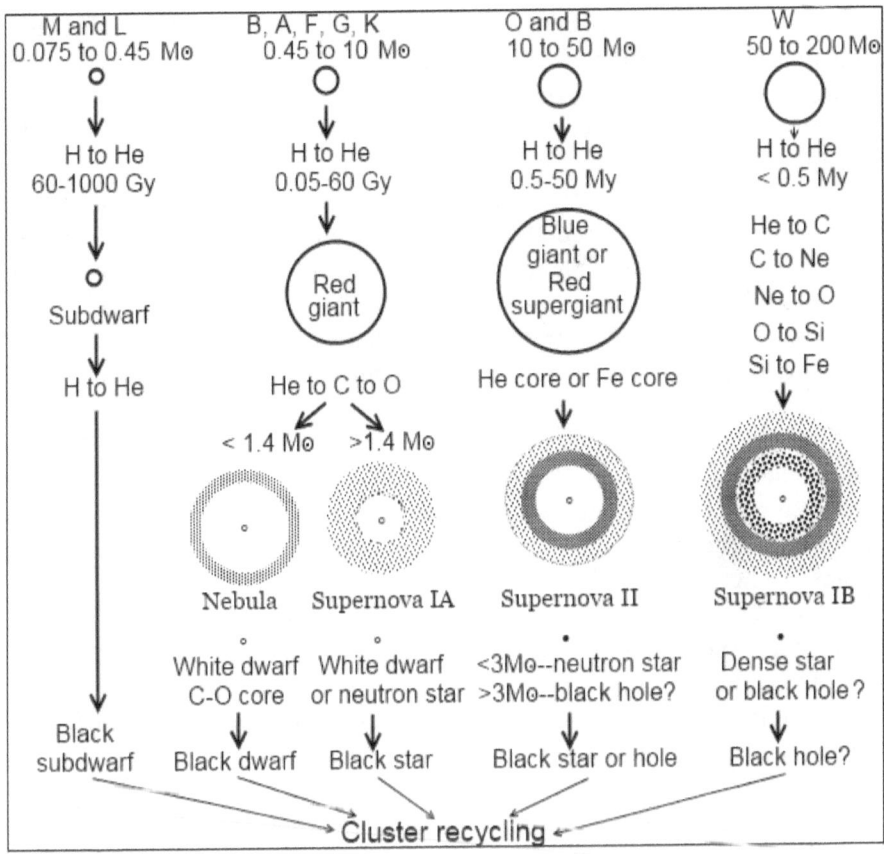

Figure 5.5 Curriculum vitae for different masses and classes of stars

Graphical curricula vitae for different masses and classes of stars are presented on figure 5.5, where the first row indicates the spectral class of the stars, followed in the second row by star masses in comparison to the sun's (M is for "mass," and the round symbol with a dot in the middle represents the sun). The circles drawn are only sketches and are not proportional to real star sizes. The next two rows show that main-sequence stars start with hydrogen-to-helium fusion that will last for a number of megayears (My) or gigayears (Gy), depending on temperature and the efficiency of the fusion cycle. The bottom two rows indicate that no matter the status of the dead star, they are all eventually recycled and crunched in a local cluster in preparation for the next maxibang.

The formation of type IA supernovae may be from accretion in a binary star system, where the larger star evolves more rapidly to the status of white dwarf and picks up material from the outer layers of its inflated red giant close companion until it reaches the Chandrasekhar critical mass, followed by rapid fusion implosion and an expanding shock wave, producing the supernova explosion. This type of supernova may be considered as a standard candle to determine cosmic distances because of a characteristic light curve whose luminosity is produced by radioactive decay of nickel to iron. But this may be questioned since it assumes that the critical mass is always exactly the same for each explosion in type IA. Some error allocation will probably have to be made for mass and surroundings variations, like companion star or neighboring dust and clouds. Other scenarios to produce type IA supernovae imply a merger of two white dwarfs or a merger of a white dwarf with any other star as long as the critical mass limit is reached.

Supernovae types IB, IC, and II are born from stars with masses of 10 solar masses and more that rapidly evolve from hydrogen burning to helium and metals up to the iron limit. When hydrogen is exhausted, the star maintains equilibrium by contracting under gravity and creates internal pressure against collapse by helium-to-carbon fusion at a temperature of about 100 million K. And when helium fusion terminates, then carbon fuses to neon at 600 million K, and the cycles of contraction-fusion continue at ever higher temperatures with neon, oxygen, and silicon until no more fusion can take place, and iron photodisintegration by gamma rays is reached around 6 billion K. The latter reaction consumes energy and will produce the central implosion and supernova.

The core may collapse at velocities of about ¼ the speed of light, with a tremendous increase in temperature at 100 billion K and density that will produce a neutron core and a neutron star with the massive escape of neutrinos. The neutron star may also recoil from a companion star and thus be spinning and moving at a higher speed than the original star. For a progenitor star of less than 20 solar masses, further collapse is halted by the degeneracy pressure of the neutrons while the shock wave carries enough energy to expel the outer layers that will form the visible supernova remnant. Above that mass, and if no higher step in degeneracy pressure is present, like quark, electroweak, or any other quanta pressure, then a black hole may result. For stars with more than 50 solar masses, a lot of work is still ongoing for stability in the models of intermediate and final steps. The extreme heat and availability of neutrons may produce most of the elements beyond iron in type II supernovae. These elements from supernovae will mix with the other elements and become the major source of heavy elements that will enrich the molecular clouds and form basic molecules and dust grains before new star formation.

5.2   Families of Galaxies and Clusters

Spiral galaxies, with their elongated arms forming flattened disks and bright central bulges often stretched in a bar shape, form the best known and most outstanding family of galaxies. Elliptical galaxies, on the other hand, look like discrete stars or lone bulges with no spectacular shape. Irregular and colliding galaxies range from the extremely complex mixture of stars, clouds, and dust to simple irregular shapes. The spiral type of galaxy represents about 35% of the total within 30 million l-y of the Milky Way while the elliptical family includes 15%, and the larger part of the pie goes to the irregular type, comprising around 50% [48].

Clusters are the locus toward which all close-by groups of galaxies, lone galaxies, and surrounding gas converge since the latter are driven by kinetic energy from expansion and by gravitational attraction. Smaller clusters are also moved by gravity and expansion energy toward the larger clusters. Thus, velocities are directly proportional to these two driving factors. Care must be taken not to interpret velocities at face value, but to take all dynamic factors into consideration and thus avoid or circumvent the dark matter model.

Important research activities are currently focusing on what kind of material is present in galaxies and clusters and how it came about [69]. What are the limits, contribution, and exchange between material phases, such as ionized, neutral, and molecular? One hypothesis includes the presence of molecular hydrogen, including ¼ of the mass in helium, everywhere in the low-density voids except in and around the filamentary structures associated with galaxy formation and clusters since that molecule is extremely stable and since EM waves may have a mean free path of many billion years in the rarefied medium between clusters and galactic walls. A more recent hypothesis is based on the analysis of quasar light that travels over large redshift z distances and in which specific absorption lines are attributed to the crossing of filamentary structure and where ionized hydrogen, including ¼ of the mass in ionized helium, would fill all the low-density voids and be invisible to EM waves since ionized atoms are a saturated plasma and nonreactive to high-energy radiation.

There are no definite answers yet, but many clues are under study in and around galactic walls and clusters. Meanwhile, to avoid the "chicken and the egg" priority problem, we may use the simplest hypothesis (Occam's razor) where we go from the original ionized atoms out of the bang to neutral hydrogen and to an expanding shell of molecular hydrogen, instead of the alternate and more complex chain of events where we would go from the ionized phase after the bang to neutral hydrogen and then back to reionized hydrogen in an expanding universe where the latter must cool down to form neutral hydrogen again and then molecular hydrogen. The latter, H_2, is mostly expected to be formed with the catalyst help of dust particles that would have been generated in a special unexplained process prior to the first star formation outburst [28].

So the formation of molecular hydrogen H_2 may proceed, in the early phase of the expanding and cooling bubble shell after the formation of some neutral hydrogen at a temperature of about 3,000 K, with the following reaction: $H + H^- \rightarrow H_2^- \rightarrow H_2 + e^-$, thus freeing an electron for the next cycle. Although there is competition for alternate reactions with the H^- anions, we know that the above reaction is probably the dominant one since the first stars are most likely formed from cold and dense molecular hydrogen clouds. When two atoms of hydrogen combine to form an H_2 molecule, then only ½ the number of particles is left, with a consequent drop in the kinetic energy and temperature of the system.

The formation of H_2 will be more important within the shell and preserved from destruction by high-energy photons since the shell is swiping a volume where a small amount of dust was left over from the previous bubbles. This dust may be a shield and a useful catalyst to produce additional H_2 from neutral hydrogen. Thus, a picture emerges, where we have an expanding shell made of molecular hydrogen clouds protected on both sides from high-energy photons emitted by neighboring active bubbles by a layer of neutral hydrogen and also on both sides by a more exposed layer of ionized hydrogen. The inside layer of ionized hydrogen and helium within the shell would then leave behind the tenuous low-density material forming the large inside volume of the bubble.

5.2.1   Spiral Galaxies and Star Globular Clusters

The spiral family of galaxies is classified according to the size of the central bulge, where *Sa* includes the largest ones; *Sb*, the medium size; and *Sc*, the smallest central bulge [76]. Moreover, when a bar is present across the central bulge, letter *B* is inserted and yields the following symbols: *SBa*, *SBb*, and *SBc*. Generally speaking, spiral galaxies with up to 200,000 l-y in diameter exhibit two or more arms forming the disk, a central bulge with or without a bar, large gas and dust clouds, the generation of large-mass and short-lived *O* and *B* stars, older stars giving a reddish core and a reddish tint to the halo on both side of the disk, and young stars giving a bluish hue to the disk.

Spiral galaxies and star globular clusters seem to have a similar origin and a different life path. Both may be formed or rejuvenated when the shell of an expanding bubble crosses an old wall or, more generally, clashes head-on with the expanding shell of another bubble.

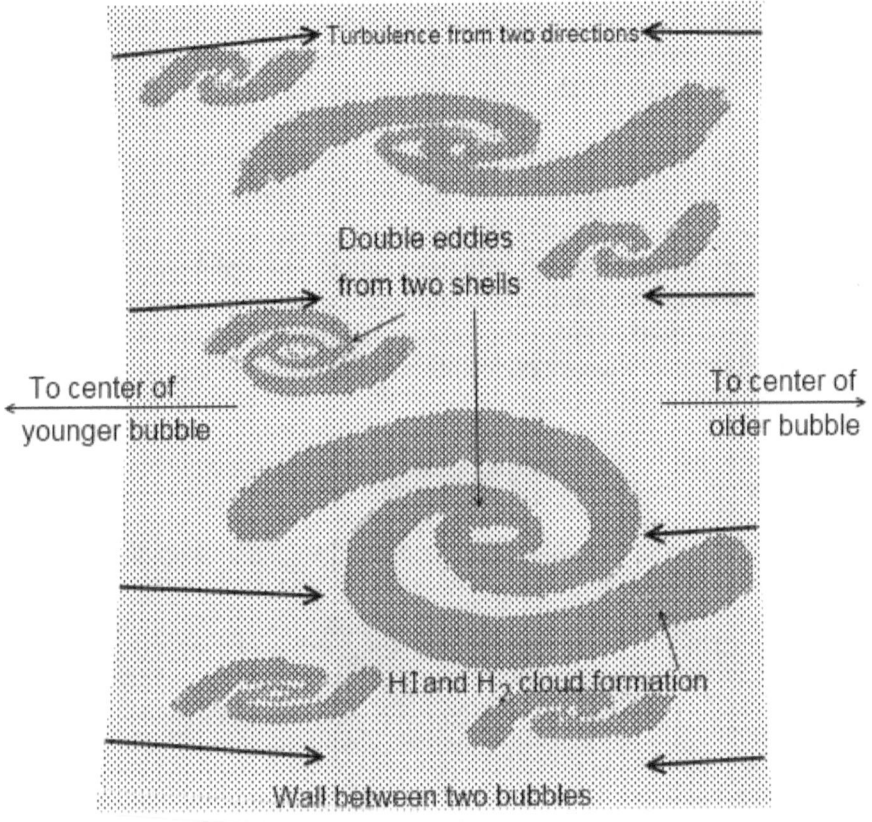

Figure 5.6 Galaxy formation between two expanding shells

Figure 5.6 is a blowup view of the formation of a group of galaxies from two expanding shells of molecular hydrogen, where the younger bubble with the strongest curvature is to the left and the older bubble to the right has a lower curvature indicated by arrows. In such a clash, differential turbulence is created from two almost opposite directions while the density of the molecular gas, always including about ¼ helium by mass, increases from the contribution of the two shells and compaction across the common wall. Note that the low-density material within the huge internal volume of bubbles is likely an ionized and invisible gas escaped from the edge of the expanding shell and representing about 90% of cosmic volume. Since the bubbles are likely filled and lined with ionized gas, we should eventually observe some signals from hydrogen—and helium-emission spectra.

Eddies with a large range of sizes start their spiraling, spinning motion, and flattening of an extended arm from the transfer of kinetic energy in the collision of the two shells. As the kinetic energy decreases, eddies will separate from the main molecular mass and generally form a more symmetric spiral with two main arms. At the center and along the arms in the disk, dense molecular hydrogen will locally spin and collapse under the force of gravity and be followed by massive star formation. Larger spiral structures will start on a first rampage mission to attract all nearby clouds and small companions and increase their elongating arms. The smallest ones will be the first to be stripped of their gaseous arms and be left with naked bulges that will become the cohort of globular star clusters and small elliptical galaxies usually accompanying the larger spiral galaxies. Midsize galaxies in the neighborhood will also be cannibalized but over a longer period of time and while star formation proceeds in protected high-density areas.

Figure 5.7 Comparison of a hurricane with a galaxy

A comparison between a hurricane and a galaxy is presented on figure 5.7, where striking similarities and differences can be observed, like fractal patterns at different scales. On the left, Hurricane Frances in the year 1996 photographed for US National Oceanic and Atmospheric Administration by a NASA satellite above Earth, and on the right, galaxy M51, also called the Whirlpool Galaxy taken by NASA's Hubble Space Telescope. Similarities include the more massive look of the spiraling arms and their widening away from the center. The radius is about 500 km for the hurricane versus 43,000 l-y (4×10^{17} km) for the galaxy. Wind speed is around 200 km/hour in a hurricane versus rotation speed of about 200 km/sec (3,600 times more) in a large galaxy. But one interesting and questionable difference is the center where a dark hole appears in the hurricane while a bright spot is observed for the galaxy. The eye of the hurricane in the center is a relatively quiet and low-pressure location toward which winds converge. It is surrounded by walls forming a cylinder where cloud density and wind speed reach a maximum. On the other hand, the center of a galaxy is a massive high-density location toward which the surrounding masses converge by gravitational attraction. It forms a spheroidal bulge where the most central part is hidden by the bright light from the bulge. So what is hiding at the center of a galaxy? How about a black hole?

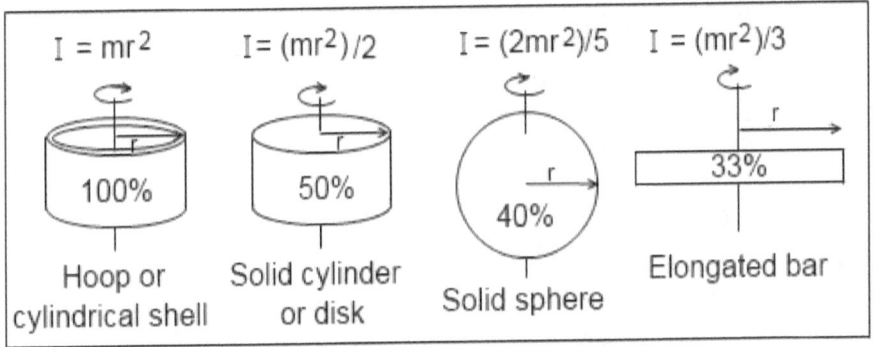

Figure 5.8 Moments of inertia from four geometrical shapes

In physics, the rotational kinetic energy about an axis is expressed with the following equation:

$$K = I\omega^2/2 \tag{5.1},$$

where K is "kinetic energy," I is "moment of inertia," and ω represents the "angular velocity" about an axis. On figure 5.8, the moments of inertia of four different geometrical shapes are drawn and identified, from left to right, with decreasing fractional values, also expressed as the percentage of a fixed common mass and corresponding to the equation above each shape.

We notice that for a fixed mass, the moment of inertia and, consequently, the kinetic energy are highest for a hoop or cylindrical shape like the one we find at the center of a hurricane or at the center of any whirling fluid. A large drop to half the kinetic energy is found if the shape changes to a solid cylinder or a flat disk like that found in star formation with planets or in a galaxy. The next step down in kinetic energy is the solid sphere, where only 40% of the energy is left, like in stars, planets, and the spheroidal bulges in galaxies. Finally, only 33% of the kinetic energy would remain for a bar-shaped structure like the one found across the center of barred galaxies. The main idea here is that not only the velocity but also the shape of a structure may be modified by kinetic energy variations.

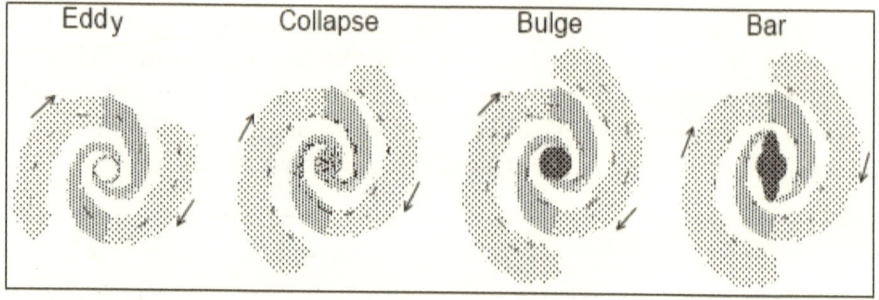

Figure 5.9 Spiral galaxy evolution from eddy to barred system

On figure 5.9, we transpose this physical information to explain the evolution of a spiral galaxy, where the collision of two shells produces an eddy with a cylindrical or ring-shaped core that will collapse to a filled cylindrical core under the pull of gravity as kinetic energy decreases and while a disk is generated from the flattening of the arms. An initial burst of star formation will produce metals from fusion and inject it more or less explosively into the arms of the galaxy to the benefit of the next generation of stars. The arms will increase in length and in mass from the input of neighboring gas clouds and smaller galaxies.

Notice that the stars, dust, and gas clouds are gradually moving in the arms toward the center of the spiral as shown by the arrows in a way similar to the clouds in the arms of a hurricane spiraling toward the central low-pressure eye. The latter movement in galaxies funnels and concentrates all the gas and dust material in the narrower portions of the arms until density and temperature in the molecular clouds are sufficient to produce new stars.

As kinetic energy keeps going down, the center of the galaxy is transformed into a larger spheroidal bulge mostly from the accumulation of stars from the first generation while the arms produce stars like our solar system and stars destined to a long life. The production of explosive stars continues at a lower and steady rate in the giant molecular clouds. Accretion of material from the surroundings is prompted by the gradual strengthening of the gravitational field in proportion to the increase in galactic mass.

Finally, as kinetic energy decreases near the center toward the lower limit associated with a bar, the velocity of the inward-spiraling material in the arms decreases to the one around the bulge, so a bar will gradually grow across the central bulge until an SB galaxy emerges. Thereafter, the galaxy will pursue its dog-eat-dog activities with the neighbors until it forms a lenticular galaxy that will be pushed and pulled into a cluster where, in turn, it will be gobbled up by a giant elliptical galaxy.

5.2.2   Lenticular and Elliptical Galaxies

Lenticular galaxies are classified as spiral-elliptical *S0* or as barred spiral-elliptical *SB0* between the most spherical elliptical ones in E7 and the largest spirals in Sa. In other words, an S0 lenticular galaxy like NGC 5866 or the large Sa Sombrero Galaxy (M104) present mixed hybrid characteristics between spiral and elliptical galaxies since a disk surrounded by an elongated or oblate spheroid of bright light is observed. Thus, large barred spiral galaxies may exhaust their supply of star-forming gas clouds during their trip toward a cluster over a period of many billion years with the ensuing consequence that their arms will also be consumed and will disappear, leaving only an elongated spheroid. As the mass and the density of the spheroid increase with mergers, we should observe an increase in kinetic energy that will transform the galaxy into a giant elliptical one with the aspect of a rounded sphere.

The main characteristics of elliptical galaxies are the presence of a reddish color produced by old stars and the absence of gaseous and dusty arms, thus hindering new *O*—and *B*-type star formation. The shapes vary from the sphere, designed type *E0*, to the most elongated cigar-shaped, designed *E7*. Giant elliptical galaxies reaching a few million light-years in diameters are by far the largest, but like for the size of fishes in the ocean, they are outnumbered by a factor of ten by dwarf elliptical galaxies.

Elliptical galaxies may be divided into two main groups: the first group involves mostly younger and smaller galaxies that have lost most of their gaseous clouds to the profit of more massive neighbors, and the second group concerns generally older and larger galaxies that have exhausted most of their star-forming clouds.

All galaxies, including the elliptical ones, are expected to hide in their center a black hole proportional in size to the mass of the galaxy so that accretion of new material should also feed the black hole. Larger elliptical galaxies are so dazzling that we cannot observe their intimate core and activities, but recent x-ray and radio wavelength data indicates that they hide one central black hole and often numerous secondary black holes in globular clusters or high-energy activities in massive stars or binary star systems.

5.2.3   Colliding and Irregular Galaxies

Irregular galaxies are characterized by the absence of a clear geometrical form where no complete ring or spiral arms may be observed. These oddities may show asymmetric or numerous bulges from mergers. They are generally rich in bright *O* and *B* young stars born from their molecular clouds.

The smaller and more numerous irregular galaxies are satellites to larger galaxies, like the two Magellanic Cloud galaxies that are gravitationally bound to our Milky Way galaxy: tidal effects produce partial or total deformation of these satellite galaxies, which end up looking like a bad dream. The larger irregular ones are the result of one or many encounters with fair—to large-size galaxies where a merger may take place or where the kinetic energy of one galaxy may sometimes, like a pendulum, be sufficient to swing back and forth in the gravitational field of the larger galaxy. Others with sufficient kinetic energy may escape the field, with damage mostly limited to the capture of gas and dust by the more massive galaxy, but in any case, a common barycenter and many secondary barycenters may be present and dictate the reshaping of the whole material. Many theoretical mathematical models have been developed on star and galactic dynamics and presented by authors such as Binney and Tremaine [4].

5.2.4   Active Galactic Nucleus or AGN Galaxies

The black holes at the center of massive stars and of all galaxies are dynamic locations where high velocities and a strong magnetic field are present and produce many different phenomena. There are recurrent time-limited EM signals associated with large galaxies. They vary from low-frequency radio waves to high-frequency x-ray and gamma ray signals.

One class of galaxies with an active nucleus is the Seyfert galaxies, which are generally found in larger spiral galaxies like NGC 1068. They show a bright compact nucleus that may imitate a quasar in visible light since the arms are occulted by the bright light. Seyfert nuclei will radiate strongly in infrared, optical, ultraviolet, x-ray wavelengths, and for about 5% of them, also in radio wavelengths. Broad emission lines very close to the center indicate star, gas, and dust velocities up to 5,000 km/sec around the compact nucleus, which is most likely a black hole.

Radio galaxies are mostly found in elliptical galaxies, like M87 at the center of the Virgo cluster, where they radiate more strongly at radio wavelength, than in spiral galaxies. They present two emitting radio lobes located around the apex of two high-velocity and opposite jets composed of energetic particles like electrons moving along helicoidal curves in a rotating magnetic field. The relativistic high-velocity electrons will first emit x-ray radiation along the jets and then emit lower-frequency radiation with typical strong radio wavelengths in the extensive lobes.

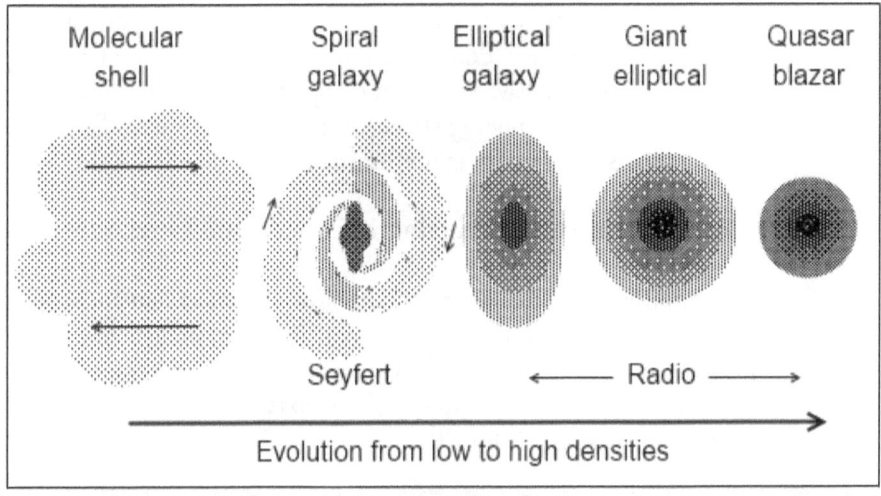

Figure 5.10 Galactic evolution from low to high density

Galactic evolution is summarized in figure 5.10, where galactic types progress in density from left to right. We start with a molecular shell expanding and clashing with a neighboring shell, where an eddy produces a spiral galaxy that evolves to a larger size and to a barred spiral where strong EM emissions from infrared to x-ray may indicate a Seyfert galaxy with an active galactic nucleus.

Then, spiral arms are gradually consumed in the intermediate step of a lenticular galaxy before emerging as an elongated elliptical galaxy where radio emission is often a witness of strong activity in the galactic nucleus, including two high-energy opposite jets and lobes. Large elliptical galaxies keep on feeding on the cluster's smaller subjects until they become giant ellipticals with periods of strong EM emission from the dense and very compact nucleus with a massive black hole at the center. Finally, when approximately one thousand galaxies and the associated gas, representing around six times the galactic star mass, have been cannibalized by a central super giant galaxy, we end up with a quasar or a blazar, depending on the orientation of the giant mass and its nucleus [53].

The hypothesis of a black hole at the center with infinite density is unacceptable from first principles like conservation of energy and space-time concepts. We know that large masses are present in massive stars or accumulate in the core of galaxies, but these are most likely stopped in their gravitational contraction at different steps as suggested by quark star models and other anticipated but not proven subquark quantum mass energy compaction schemes or by the presence of some superluminal black energy. With this in mind, a black hole is an extremely dense mass located at the center of a star or galaxy from which no light or EM wave can escape but with no influence on the gravitational field other than being stronger, more concentrated. The latter field escapes unscathed since gravity is not limited by the speed of light.

5.2.5 Intra—and Perigalactic Clouds and Dust

With the extreme galactic variation in size, location, and dynamic interactions as presented in hundreds of written articles and images every year, we will only summarize some of the galactic characteristics and the extent of gas clouds and dust in and around solitary field galaxies and groups of galaxies located in the walls between bubbles [9].

Cosmic galactic group parameters:			
Physical parameter	Number	Units	Remarks
number of groups	6.9848E+11	groups	if 200groups/bubble
average radius	1.50	Mpc	megaparsec
average radius	4.89E+06	l-y	light-years
average radius	4.6308E+22	m	
average volume	4.1597E+68	m³	cubic meters
average mass	1.39E+42	kg	
average mass	8.32E+68	Mp	proton masses
average density	2.00	Mp/ m³	proton mass/cubic meter
nb large galaxies	2	galaxies	1.6E+11 stars each
nb small galaxies	38	galaxies	1.0E+10 stars each
nb galaxies/group	40	galaxies	2 to 50 galaxies
nb of stars/group	6.98E+11	stars	sun masses
nb of stars/galaxy	1.75E+10	stars	sun masses
mass of stars/galaxy	3.47E+40	kg (sun masses)	intragas included
mass of stars/group	1.39E+42	kg	sun masses

Table 5.2 Cosmic galactic group parameters

In the cosmic parameters for galactic groups in table 5.2, we assume a ratio of two hundred groups per bubble, which is an average since each bubble shares common walls with its neighbors according to its history of expansion and encounters. The average spherical radius is only indicative since most groups are expected to inhabit generally flattened 2-D walls. Our own group serves as the main model for the number and distribution of large and small galaxies, and our sun serves as the model for star mass although we know that red dwarf M stars are by far the most numerous, with an average mass about $^1/_3$ of that of the sun. So we find that galactic groups have an average density of two proton masses per cubic meter, which is about ten times more than the cosmic average. The number of stars per galaxy is the equivalent of a low average value of 17.5 billion solar masses since most galaxies are relatively small compared to large galaxies, such as the Milky Way.

When a galaxy is observed with a large range of wavelengths, a whole new world is discovered. For example, the Southern Pinwheel Galaxy, M83, shows wide and long arms of neutral hydrogen when observed at the 21 cm wavelength with a radio telescope. These extraneous arms, not visible at optical wavelength, spiral and extend four to six galactic radius all around and away from the central visible part. They can be compared to unspooling fluffy ribbons originating from the torn-apart neighboring galaxies in the process of being cannibalized by M83. In the ultraviolet-wavelength telescope, these arms look like a galactic skeleton since the energetic photons are emitted by young stars born out of the limited molecular clouds concentrated within the thick protective shield of neutral hydrogen and dust. A similar phenomenon is present to a lesser extent around Andromeda M31, around our Milky Way galaxy, and more or less around all cannibalizing galaxies.

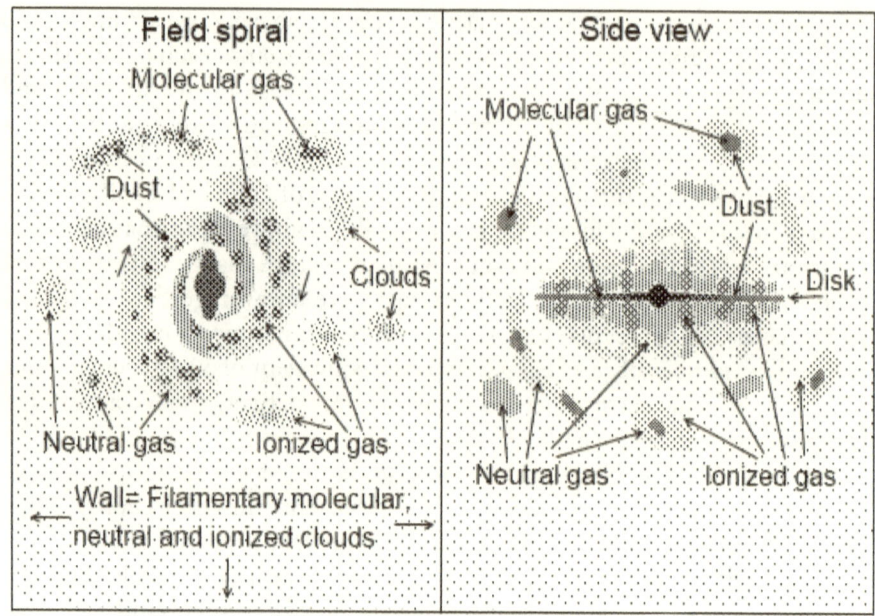

Figure 5.11 Field spiral and surroundings in a wall

The drawn sketch represents a barred spiral galaxy, which is typical of large galaxies located in a squeezed wall between bubbles. On figure 5.11, we show a face view on the left and a side view of the galaxy to the right. The wall, mostly made on both sides of ionized gas from two shells exposed to high-energy photons, includes a mixture of neutral gas and molecular hydrogen where the gas is protected from ultraviolet and high-energy photons toward the center. It also includes dust where dense, hot stars have released their metals within a cloud.

The arms of the galaxy are the location where star formation takes place from the giant molecular clouds that are gradually funneled and spiraling in toward the center. It is also the site where supernovae expulse at high velocities their external layers containing most metals that will combine to form dust and larger molecular particles. Part of this material involved in the implosion/explosion reach a sufficient velocity to escape from the central plane of the galaxy and reach a distance of up to a few kiloparsecs and then either escape from the galactic gravitational field or slow down and fall back like a fountain.

A lot of activity is associated with the end of a bar, when present, such as intense star formation where stars are integrated to the particular kinetic energy and velocity of the bar and where an important outward wind may blow particles away from the high star density and form clouds above the disk.

The bulge usually contains older, more mature stars and very little molecular gas, so star formation is halted near the galactic center, but a large amount of ionized gas from the stars is present and steadily accreted to the central black hole [5].

As we move away from the close-by galactic corona for which most of the material may come from the galaxy itself, we find clouds and filaments, as in the Southern Pinwheel Galaxy M83, that have been pulled from neighboring and usually smaller galaxies [34]. This gas and dust material may contain an external envelope of gas ionized by high-energy photons originating from the galaxy. The interior of these stolen clouds may contain protected molecular hydrogen surrounded by neutral hydrogen and dust. In merging situations, a filament of dense and cold 10 K or so molecular hydrogen may produce by compression a series of starbursts, visible in ultraviolet and sometimes in optical observations, during its trip to the more massive galaxy.

5.2.6   Clusters of Galaxies and Intercluster Medium

As explained previously, galaxies are born and generally lined up in the plane of the wall formed by two colliding shells from two expanding bubbles. While they are gradually displaced sideways across the wall by the kinetic energy of the expanding bubble, they are transformed to cannibalizing spirals that will grow in size, mass, and density until they move out of the wall into the concave triangular section toward the closest tetrahedral cluster. Then, they will mute to S0 and to elliptical galaxies since they are cut off from their fresh supply of cold and dense gas. The next mutation step brings them in the stomach of a giant elliptical galaxy with some of the surrounding hot compressed gas falling in. Finally, a quasar will form in a large hexahedral cluster.

Cosmic cluster parameters:			
Physical parameter	Number	Units	Remarks
number of clusters	6.9848E+09	average clusters	500 galaxies/cluster
average radius	4.10	Mpc	1.5 to 10 Mpc
average radius	1.34E+07	l-y	light-years
average radius	1.2658E+23	M	
average volume	8.4946E+69	m^3	cubic meters
average mass	1.70E+44	Kg	
average mass	1.02E+71	Mp	proton masses
average density	12,00	Mp/m^3	proton mass/cubic meter

nb large galaxies	50	galaxies	1.6E+11 stars each
nb small galaxies	450	galaxies	1.0E+10 stars each
nb galaxies/cluster	500	galaxies	50 to 1,000 galaxies
nb of stars/galaxy	2.50E+10	Stars	sun masses
mass of sun	1.99E+30	Kg	
star mass/galaxy	4.98E+40	Kg	average
star mass/cluster	2.49E+43	Kg	
gas mass/cluster	1.45E+44	Kg	
ratio gas/star mass	5.84		about 6× more gas
Parameters for invisible gas in walls:			
wall gas volume	6.90E+80	m³	out of groups + clusters
wall gas mass	1.12E+54	Kg	
wall gas mass	6.73E+80	Mp	proton masses
wall gas density	0.97	Mp/m³	proton mass/cubic meter

Table 5.3 Average physical parameters for cosmic clusters and gas in walls

The upper part in table 5.3 is showing cosmic-scale parameters for the average-size cluster, which contains about five hundred galaxies. Note that we have for each bubble the equivalent of two average-size tetrahedral clusters. After mergers, this will yield the large hexahedral-size cluster, including the mass of one thousand galaxies plus the equivalent gas mass of six thousand galaxies that will be compacted and produce the next bubble cycle. The average mass of one cluster is about 1.65 times the mass of one bubble, but its volume is more than three hundred times less, so cluster density is more than five hundred times that of bubbles. A cluster ratio of six times more gas than star mass has been applied in order to account for the large amount of high-energy gas detected within clusters. The last four lines of table 5.3 show the parameters for the generally invisible gas present in all the walls between active bubbles and outside of group and cluster volumes. This is not related to fossil walls within bubbles. The volumetric density of this gas is about five times more than the average cosmic density.

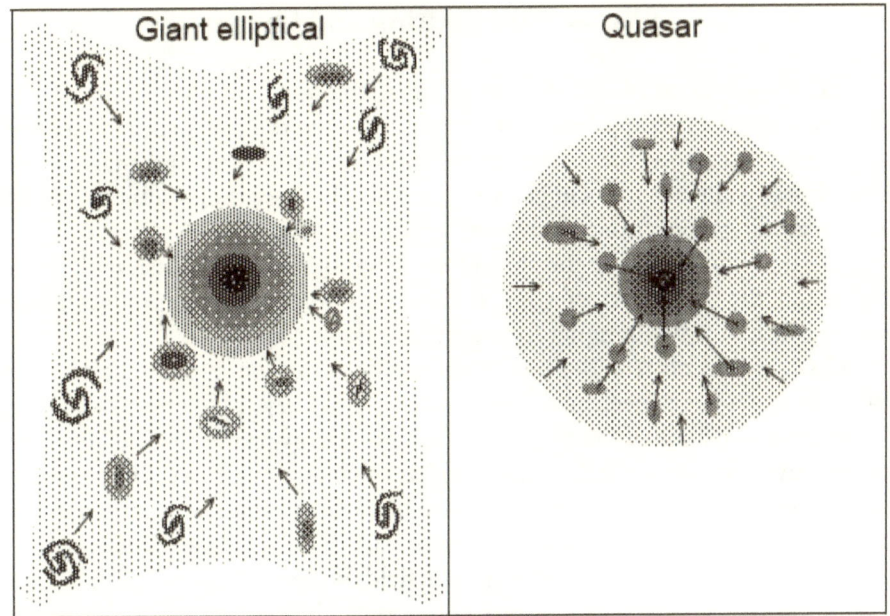

Figure 5.12 Cluster with a giant elliptical galaxy and a quasar

On figure 5.12, we illustrate a giant elliptical galaxy on the left and a quasar on the right, both located within a cluster of galaxies. The giant elliptical cluster is spherical since it involves a very large mass with increasing density toward the center, like in a star or a large planet. Spiral and elliptical galaxies formed in the walls are moving in the cluster via the concave triangular section between three bubbles, with all their surrounding gas, dust, and debris, including stars, planets, and satellites. The latter material is compressed and heated to a high temperature while galaxies are merging and falling in toward the central ogre from which a double jet of particles is emerging at high velocity and signaling the presence of an insatiable central black hole.

The quasar is the last and final phase in galactic evolution when the equivalent of a critical mass between five and ten thousand galaxies has been gathered in a giant hexahedral cluster, including mostly hot gas and the smaller contribution of about one thousand galaxies. The cluster is gradually compressed and partly cooled by the emission of a large range of EM waves carrying some heat away. The last galaxies and the rest of the gas are rushed in toward the extreme black hole in the center over a period of a few hundred million years. In the end, like the phoenix consumed in a fire and reborn from its ashes, the quasar brings the whole cluster to a final contraction before its rebirth in a maxibang and expansion as a new local bubble.

5.3 Quasar Distribution and Dynamics

When we observe galaxies or quasars, the total number available for observation increases approximately in proportion to the distance, up to a limit where they start to decrease because absorption, extinction, and masking by intervening walls, clouds, and galaxies gradually hamper their visibility in EM wave recordings by various land-based or orbital telescopes.

With the expansion of the shell of every bubble come the clash with other shells and the formation of stars and galaxies where heat is produced and released as EM waves ranging from gamma rays to radio.

Quasars and blazars, depending on their orientation, are known to emit strongly over a large spectrum of frequencies [73]. They are likely the source of acceleration for some of the most extreme high-energy proton and helium-core cosmic rays. Two opposite and powerful jets of charged particles, like electrons and protons, are observed from the active black hole center of a large number of quasars. These jets represent the visible part of the particles accelerated to relativistic speed in the gyrating magnetic field of the rotating hooplike central mass from which nonthermal x-rays and optical photons are emitted in a way similar to synchrotron emission in a lab. At the end of the jets, a plume or a cloud of EM waves ranging from visible to radio is observed. Digesting all these galaxies and gas entrains some spectacular fountains of light since the energy output of a quasar is comparable to the luminosity of hundreds and even thousands of galaxies.

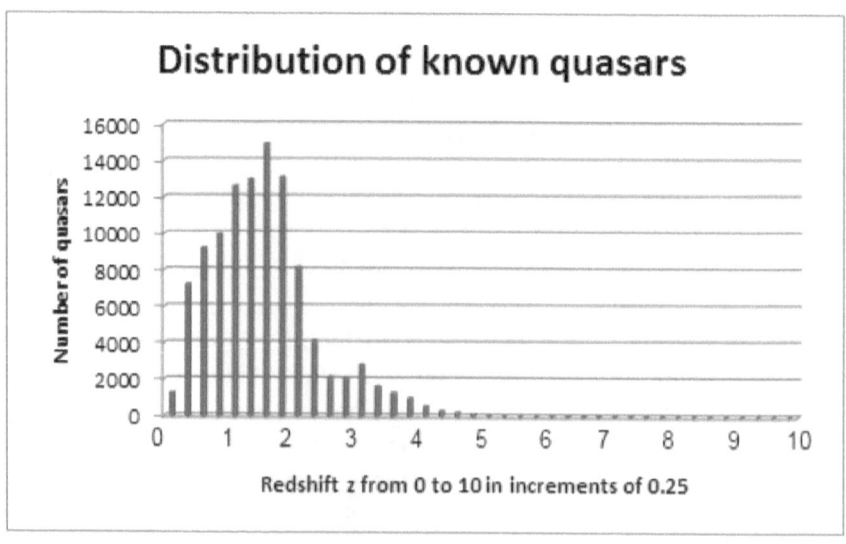

Figure 5.13 Distribution of known quasars from the Vizier catalogue

When we take the Vizier catalog, the Schneider 2010 SDSS quasar catalog where 105,842 quasars are presented, and plot the redshifts within interval increments of z = 0.25 versus the number of quasars, we obtain figure 5.13. For the first two values, the number of quasars increases very rapidly and is followed by a steady but less important increase up to z = 1.75, where a maximum is reached in the interval with about fifteen thousand quasars. Then, a rapid decrease is observed up to z = 5.5, where absorption destroys most EM waves coming to us from quasars located afar. A little kink is seen around z = 3 and is probably related to irregular volume coverage.

Since we expect an average cosmic radius in the order of z = 10 and since absorption and redshifts are the main handicaps, we should be able to record the presence of quasars up to that redshift by observation of the whole infrared to radio range of frequencies with more sensitive and accurate instruments than those presently available. Absorption of ultraviolet photons by neutral hydrogen across the Lyman-alpha forest and of high-energy wavelengths by a variety of absorbers present in the expanding shells, walls, and clouds located between us and the quasars is a gradual phenomenon increasing with the number and intensity of encounters along the line of sight until no photons are left to be absorbed [67].

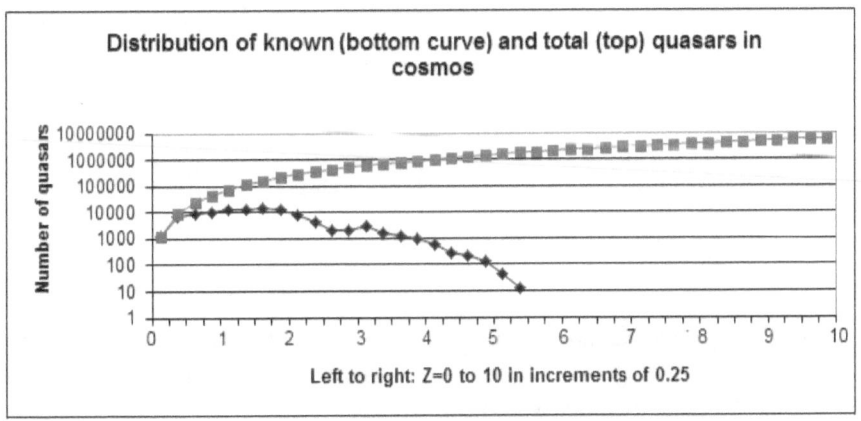

Figure 5.14 Known and projected distribution of quasars

The bottom curve of figure 5.14 is a repeat of figure 5.13 at a logarithmic scale, and the top curve is a volumetric projection in increments of z = 0.25 of the total number of quasars that we may expect to discover with improved instruments and time. Of course, the projection is an initial estimate where the last interval of z = 0.25 contains over 6 million quasars. The constant total cumulative number at any epoch would then be over 83 million quasars in the cosmos, so the number discovered (about 105,000 and growing) represents only about 0.13% of the total. Also note that these are spherical projections to obtain a total value although we are not located at the center of such a sphere.

Quasar and maxibang parameters			
Physical parameter	Number	Units	Remarks
mass of sun	1.99E+30	kg	
number of quasars	8.30E+07	quasars	Projected
bubble/quasar	4.2077E+01	ratio	
ave. bubble radius	8.6177E+20	km	
ave. expansion vel.	4,440	km/sec	across a bubble
ave. radial exp. vel.	2,220	km/sec	
radial expan. time	3.8818E+17	seconds	
ave. bubble lifetime	1.2301E+10	years	
ave. quasar lifetime	2.9235E+08	years	
quasar birth/death	3.52	years	average
quasars discovered	1.05E+05	quasars	Vizier/Schneider 2010
ratio total/discovered	7.90E+02	ratio	for quasars
probab. to observe	1 in 2,780 years	probability	for discovered quasars
nb clusters in bang	1	cluster	1 large cluster/bubble
nb galaxies in bang	1,000	galaxies	1 hexahedron with gas
ave. mass in bang	3.40E+44	kg	1 large cluster
sun masses in bang	1.71E+14	solar masses	or 171 trillion suns

Table 5.4 Parameters for quasars and maxibangs

With the projected number of quasars and the ratio bubble/quasar in table 5.4, we find the average bubble lifetime from the expansion velocity and also the average quasar lifetime, which implies that a quasar is born in the whole cosmos about every 3.5 years and that one implodes to a maxibang in that same period of time. But since only about one out of eight hundred quasars have been discovered, then our chance to observe a new quasar or a maxibang is about 1 in 2,800 years if we keep observing that same volume for that period of time. Therefore, to observe the birth or the death of a quasar in a human lifetime would require the steady observation of a much larger cosmic volume. The staggering amount of 171 trillion solar masses equivalent is involved in the triggering of a maxibang event. It includes the mass of one thousand galaxies and six times as much in gas contribution.

6. DYNAMICS OF GALAXIES AND CLUSTERS OF GALAXIES

Galactic rotation velocities, measured from Doppler shifts and distances to center of mass, present flat curves away from the center and up to the edge of most large spiral galaxies. A similar problem is found in clusters of galaxies where apparent orbiting or infalling galaxies present excessive redshift velocities incompatible with dynamically stable systems. Cumulative masses, calculated near the edge of these galaxies from flat rotation curves, exceed what is expected from Kepler-Newton dynamics but can be accounted for if corrected gravitational forces are computed and applied to rotation velocities of galactic disks. Kepler-Newton laws apply only for idealized point-like masses or uniform-density spherical masses where force decreases as the inverse square law and where the geometric center and the center of mass coincide. In nature, all masses have dimensions, and none is perfect since field frictions are present everywhere [39].

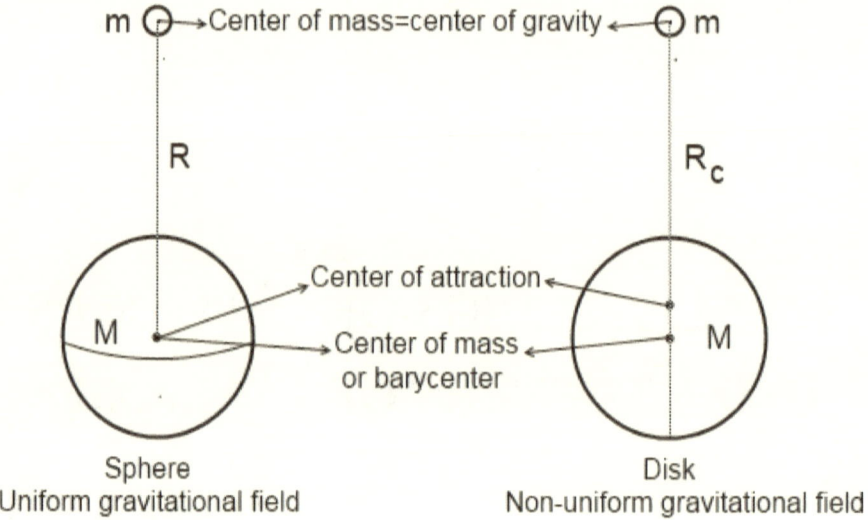

Figure 6.1 Comparison of gravitational attraction from a sphere and from a disk

A comparison of the gravitational attraction from a sphere and from a disk is presented on figure 6.1. On the left, a sphere with a large mass M generates a uniform gravitational field in all directions, so for a small body with point-like unit mass m, the center of mass and the center of gravity are located at the same point. Since m is negligible, the center of mass of the sphere and the barycenter of the system are located at the center of the large sphere, where the center of attraction is also found. The center of attraction is defined as the point toward which bodies tend by gravity. The term *center of attraction* is used for a clear distinction from the term *center of gravity* that is generally related to the average location of the weight of a body or object in a gravitational field.

But for a homogeneous disk made up of a large number of separate elements, as shown on the right, the gravitational field is not uniform in all directions although the center of mass of the disk and the barycenter of the system are located at the center of the disk. For a uniform disk, the center of attraction is located on the disk and along the line across the center of the two bodies. The center of attraction is also a point about which a central force acts and may produce a torque. R_c represents the distance between the small orbiting mass m and the center of attraction located on the disk. The orbiting mass is rotating about the point at the center of attraction while some separate bodies within the disk and closer to the center may simultaneously be orbiting about the center of mass with a zero torque.

The force is zero everywhere inside a perfectly uniform shell. In this section, we show that for a uniform-density disk and except at the geometric mass center, the force is not zero inside a ring, and larger than the inverse square law away from the edge of the disk. Corrected velocities and also equivalent Keplerian velocities, unlike the apparent Doppler shift velocities, will yield corrected masses corresponding to the volume density of physical entities, like galaxies, and thus will be more compatible with other methods of mass evaluation like mass/luminosity distributions. No need to modify classical dynamics or to account for unknown dark matter [40].

The model will be expanded to a stack of disks and applied to four typical concrete examples, which are the large galaxy NGC 3198, with a radius of 30 kpc; the small galaxy NGC 4062, with a radius of 4 kpc; the barred galaxy NGC 4389, with a radius of 6 kpc; and the elliptical galaxy NGC 3379, with a radius of 12 kpc. Then, the results will be extended to a more statistical number of forty-six galaxies in order to draw some significant features, for example, about low—and high-surface brightness galaxies. A linear string of masses will also be discussed to understand the more elongated filamentary structures.

6.1   Disk Model and Equations

In the absence of an ideally large three-dimensional database to study the gravitational relationships between billions of masses with complex orbits in spiral galaxies, a simplified model comparing a galaxy to a disk or a stack of disks is presented.

The relationship, between a unit test mass located at the edge of a disk and this same disk divided in a sequence of radiuses and a set of angles, is discussed in terms of gravitational forces, velocities, masses, and ratios. The test mass will also be gradually moved across the disk and away from the edge of the disk to measure the importance of velocity anomalies and their effects on mass computations.

Equations related to the disk model are presented. Then, graphs and results are discussed until the method is demonstrated. The few examples presented, including galaxy NGC 3198, will be computed with a stack of up to thirty disks to account for density variations across the galactic disk.

All gravity-related equations applied are from the Kepler-Newton classical physics and without modifications. Another set of equations, related to geometry and trigonometry between unit mass and disk, is used and developed here to measure total gravitational force from the total mass and cosine-corrected gravitational force from the truncated rings divided in point-like masses. Finally, matrices are presented to compute any number of *m* rows (rings) by *n* columns (disks) in practical applications to galaxies. Other methods of summation may be used to obtain similar results.

6.1.1   Disk Model Presentation

In order to compute the total and the cosine-corrected gravitational force between a unit test mass and a disk, the following geometric model is used with appropriate symbols.

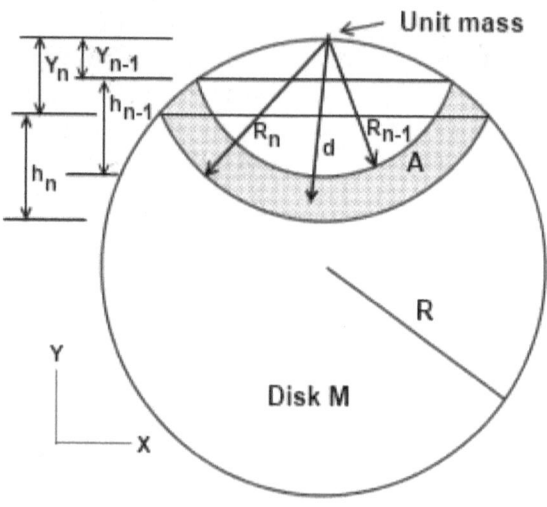

Figure 6.2 Disk *M* under study with segment divisions

Disk *M* is shown on figure 6.2 with a horizontal x-axis and a vertical y-axis. The unit mass on the edge of the disk is the center from which R_n variable radius values are generated to cover the whole disk with thin, truncated circular ring areas like *A* (in gray). Since the distance between the unit mass and any ring is constant, then the force between the unit mass and equal masses in the ring is constant, and the ring may be divided in equal point-like masses. Distance *d*, in the middle of the ring, is midway between any R_n and R_{n-1} values. Radius *R* is the fixed radius of disk *M* under study. The upper part of figure 6.2 is reproduced in figure 6.3 to show areas added and subtracted to obtain the truncated ring area *A*.

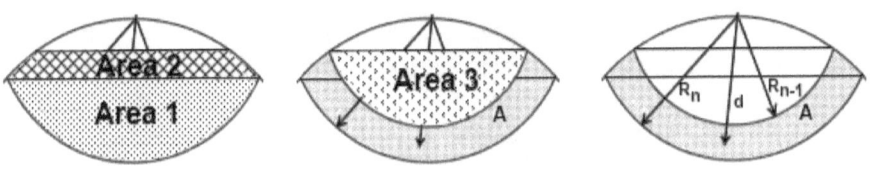

Figure 6.3 Truncated ring computations

Area A is the truncated part of a flat ring in a disk and equal to the segment with height h_n (area 1), plus the flat area between the two chords with height Y_n and Y_{n-1} (area 2), and minus the segment with height h_{n-1} (area 3). The idea is to use equidistant R_n values and associated masses to compute the gravitational force between the test mass of one unit and every ring to obtain total gravitational force values F_t for each ring and ΣF_t for the whole disk. At the same time, cosine-corrected gravitational force values F_c from every ring and ΣF_c for the disk will be computed.

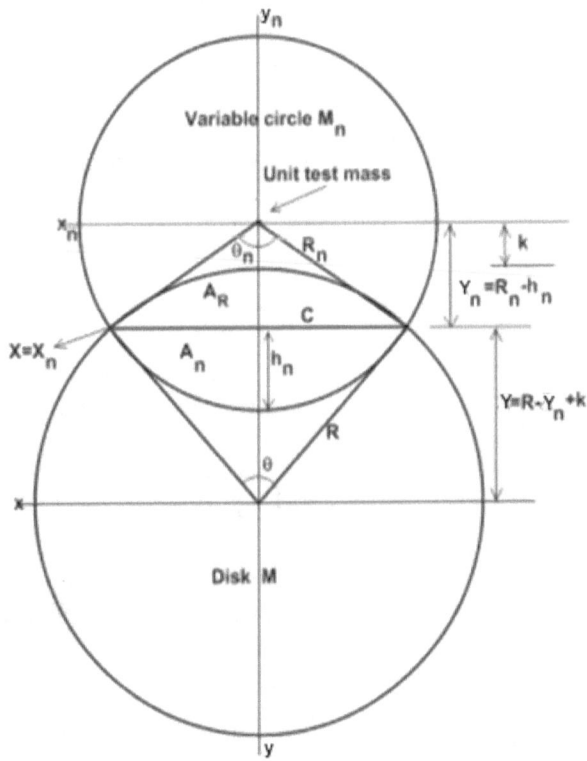

Figure 6.4 Disk M and variable circle M_n with associated symbols

Variable circle M_n is presented on figure 6.4 at a distance R_n from the unit test mass, with its horizontal x_n axis and its vertical y_n axis. Angle θ_n is formed between any two R_n radiuses running from the origin of circle M_n to the extremities of chord C. Likewise, angle θ is between two R radiuses and runs from the origin of disk M to the end of chord C. Area A_n with height h_n is a part of circle M_n limited by chord C. Area A_R is a part of disk M also limited by chord C.

The height of k represents the distance between the unit test mass and the edge of disk M. When k = 0, the test mass is located right on the edge of disk M, as shown on figure 6.2. When k is positive, as on figure 6.4, the test mass is moving away from the edge, and with k negative, the test mass is moving across the disk toward the center.

6.1.2   Circle and Disk Equations

From the graph of figure 6.4, we have
$$Y = R - Y_n + k \tag{6.1}$$
and
$$Y_n = R_n - h_n \tag{6.2}.$$
The extremities of chord C are located at the junction of disk M with circle M_n where $X = X_n$.

From the equation of the circle, $X^2 + Y^2 = R^2$ and $X_n^2 + Y_n^2 = R_n^2$, and from equations 6.1 and 6.2, we have
$$R_n^2 - Y_n^2 = R^2 - Y^2 \tag{6.3}.$$
Replacing Y, we find
$$Y_n = (R_n^2 + 2Rk + k^2)/(2R + 2k) \tag{6.4}.$$
And replacing Y_n gives
$$h_n = R_n - ((R_n^2 + 2Rk + k^2)/(2R + 2k)) \tag{6.5}.$$
Area A of the truncated ring is explained in figure 6.2 and combined to figure 6.4 to obtain
$$A = A_n - A_{n-1} + (A_R - A_{R-1}) \tag{6.6},$$
where A_{n-1} and A_{R-1} are the preceding values of A_n and A_R, respectively, and correspond to radius R_{n-1}.

Available circle formulas may be combined to the above geometry and to trigonometry to find angles, chord lengths, sector, and segment areas as follows.

Sector area for circle M_n is found from
$$\text{sector area} = R_n^2(\theta_n/2) \tag{6.7}$$
and for disk M from
$$\text{sector area} = R^2(\theta/2) \tag{6.8},$$
where the angle for M_n is

$$\theta_n = 2 \arccos(Y_n/R_n) \tag{6.9}$$

and for disk M

$$\theta = 2 \arccos(Y/R) \tag{6.10}.$$

The area of the two triangles above chord C on figure 6.4 is
$$M_n \text{ triangles} = Y_n(C/2) \tag{6.11}$$

and for the two triangles below chord C:
$$M \text{ triangles} = Y(C/2) \tag{6.12},$$

where $C/2$ is defined as
$$C/2 = R_n \sin(\theta_n/2) \tag{6.13}$$
or
$$C/2 = (R_n{}^2 - Y_n{}^2)^{\frac{1}{2}} \tag{6.14}.$$
Then, area A_n in circle M_n is equal to the sector area, equation 6.7, minus the two triangle areas, equation 6.11, combined with equation 6.14:

$$A_n = R_n{}^2(\theta_n/2) - (Y_n(R_n{}^2 - Y_n{}^2)^{\frac{1}{2}}) \tag{6.15}.$$

And area A_R in disk M is similarly expressed as

$$A_R = R^2(\theta/2) - (Y(R_n{}^2 - Y_n{}^2)^{\frac{1}{2}}) \tag{6.16}.$$

Equations 6.15 and 6.16 and also equivalent equations for A_{n-1} and A_{r-1} will be introduced in equation 6.6 to determine area A of the truncated ring for each R_n radius value across the entire disk:

$$A = [R_n{}^2(\theta_n/2) - (Y_n(R_n{}^2 - Y_n{}^2)^{\frac{1}{2}})] - [R_{n-1}{}^2(\theta_{n-1}/2) - (Y_{n-1}(R_{n-1}{}^2 - Y_{n-1}{}^2)^{\frac{1}{2}})]$$
$$+ [(R^2(\theta/2) - (Y(R_n{}^2 - Y_n{}^2)^{\frac{1}{2}})) - (R^2(\theta_{-1}/2) - (Y_{-1}(R_{n-1}{}^2 - Y_{n-1}{}^2)^{\frac{1}{2}}))] \tag{6.17},$$

where subscript −1 indicates the previous value.

The sum of all the ring areas, ΣA, is the surface of the disk. The volume and the mass of the disk are of the same value as the surface since a thickness of one unit is used as well as a density of one unit to simplify the model. Furthermore, the gravitational constant, G, may be left aside since it does not affect ratios in the model and it can be reintroduced at will in concrete examples. The purpose is to find ratios to evaluate velocity and mass relationships between a test mass of one unit and the sum of the rings of increasing radiuses across a whole disk and then transpose the results to galactic scale. In the model, the ratio between the radius and the thickness of the disk varies from 1 to 150 because most galaxies seem to have a ratio radius/thickness less than 150.

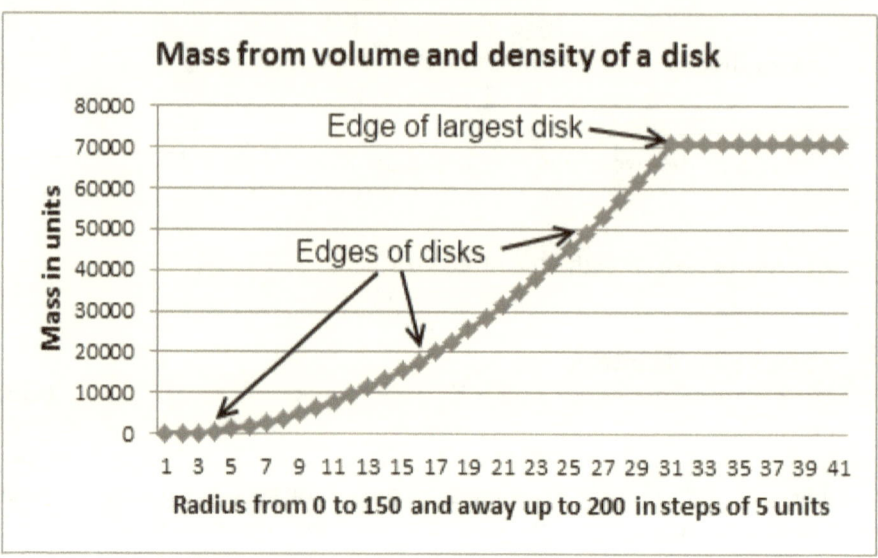

Figure 6.5 Mass computed from volume and density at the edge of disks increasing in size and away from the edge of the largest disk

The curve on figure 6.5 shows the computed increasing mass values at the edge of disks increasing in size from a radius R = 0 on the left to R = 150 distance units on the right, where it flattens out away from the edge of the largest disk over a distance of fifty units for a total distance of two hundred units. Values, starting at zero, are computed every five distance units for a total of forty-one computations across the horizontal scale. Mass is additive and increases with disk radiuses until it flattens out to a maximum at the edge of the largest disk. It is expressed in units on the vertical scale. This calibrated mass curve is calculated with the volume-density equation

$$M = \pi R^2 t_d \rho \qquad (6.18),$$

where R = 0 to 150 is disk radius, t_d = 1 is the thickness of the disk, and ρ = 1 is the density. So in this model, masses are simply equal to the area of the disks.

Two kinds of gravitational forces, three different velocities, and two radial distances associated with the mass distribution on figure 6.5 will be studied.

6.1.3   Cosine Corrected Gravitational Force

Equation 6.17 is applied to every truncated ring in the disk to obtain separate area, volume, and mass for the rings. Total gravitational force for each ring is then computed from Newton's equation,

$$F_t = GMm/d^2 \qquad (6.19),$$

where F_t is the total gravitational force per ring, G is the gravitational constant, M is the mass of the disk per ring, m is the test mass equal to one, and d is

$$d = (R_n + R_{n-1})/2 \qquad (6.20).$$

Errors from averaging d in the middle of R_n intervals are proportionately lower when a larger number of R_n values are used.

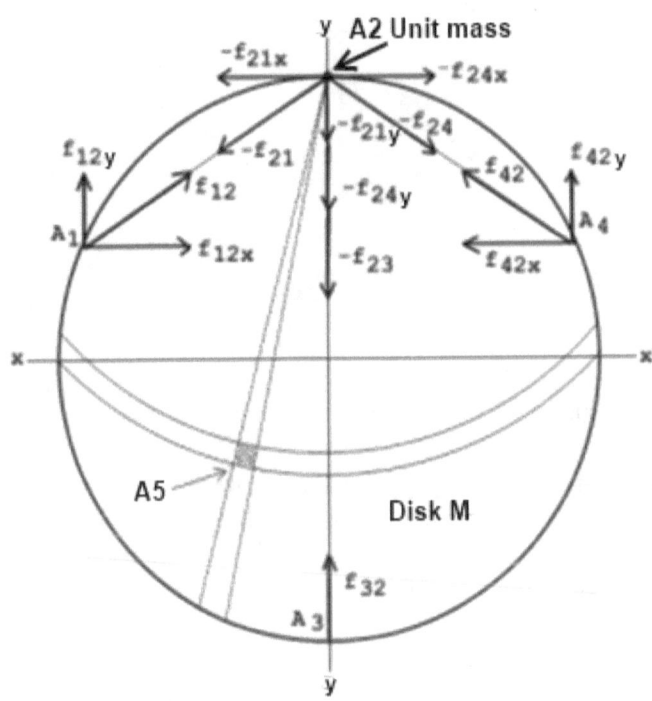

Figure 6.6 Cosine angles across the rings and vectors

One truncated ring, formed by the interval between two sequential R_n values, is reproduced on figure 6.6. It is intersected at $A5$ by a small portion of θ_n angle. Data sampling is accomplished by dividing the whole disk in small point-like masses. The size of data samples for a fixed angle increases away from the unit mass with distance. Figure 6.6 also shows a graphical example of the vector aspect of gravitational force. When unit mass is located at point $A2$, it attracts each of the three point-like unit masses located at A_1, A_3, and A_4, as shown with the vectors and their x and y components.

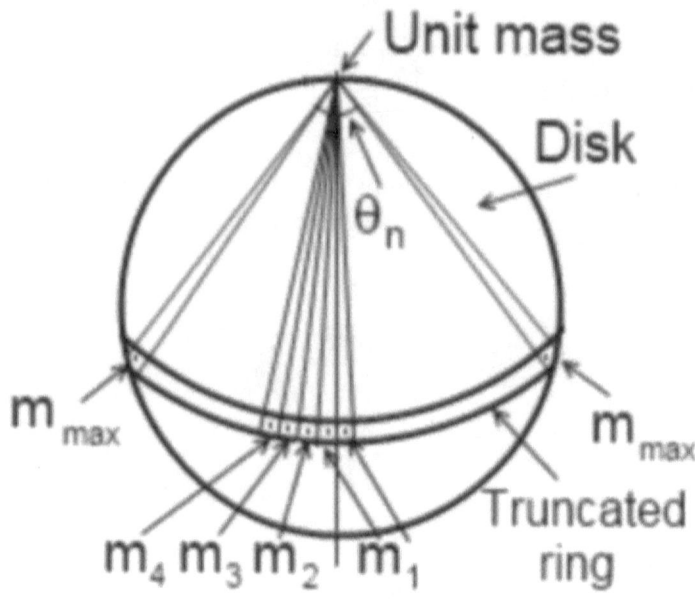

Figure 6.7 Cosine summation for one truncated ring

As shown on figure 6.7, all truncated rings are divided in small portions, m_1, m_2, . . ., up to m_{max}, by a large number of equally spaced angle subdivisions that may be selected arbitrarily in proportion with the number of R_n subdivisions. Angle θ_n has a maximum of 180 degrees or π radians, and its subdivisions are symmetric on each side of the y-axis. Force vector values will be found from the cosine of each subangle. Since mass is equally subdivided within every ring, it is possible to sum all the cosine values from subangles crossing a ring with the following equation,

$$\text{Cosine sum} = \sum_{m=1}^{m_{max}} \cos[\theta_n(((2m) - 1) / (4m_{max}))] \qquad (6.21),$$

and obtain corrected gravitational force, F_c, for that ring if we multiply the cosine sum by the point-like mass within one subangle. Then, the sum of F_c force vector values for all rings, ΣF_c, will give the cosine corrected force for the whole disk.

6.1.4   Graphs and Results

Most of the graphs and results will be for disks in the range R_n = 0 to 150 units, which include radius/thickness ratios similar to those for a majority of regular spiral galaxies. But an additional distance, ranging from k = 5 to 50 units, will be presented to show what occurs away from the edge of the disk. Negative k values will also be explored across the disks. A total distance of 200 units in steps of 5 units will yield forty-one values in the following theoretical model.

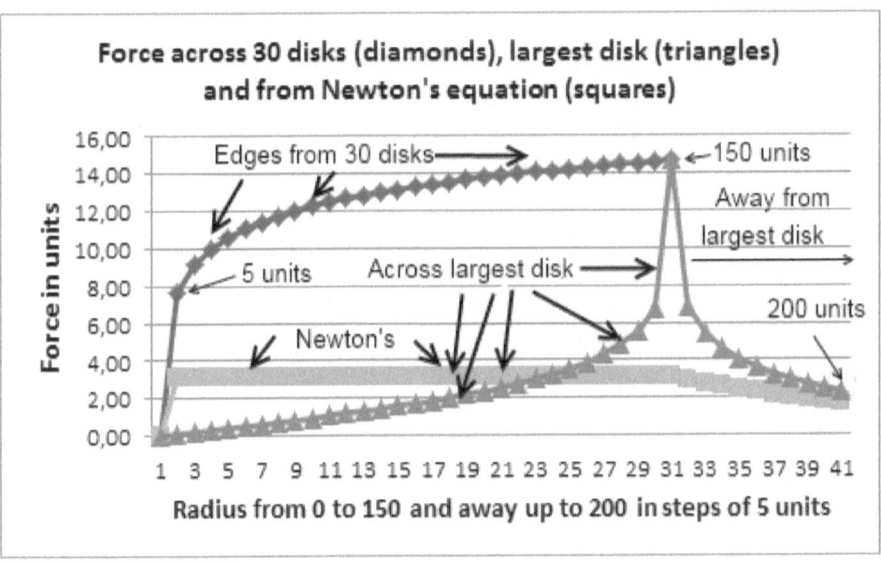

Figure 6.8 Comparison between cosine-corrected gravitational force F_c across thirty disks. Also, F_c across and away from the largest disk with a radius of 150 units (triangles) and Newton's gravitational force F_n (squares).

The horizontal scale is the same as on figure 6.5, and the vertical scale shows gravitational force in relative units. The curve with diamonds represents net force vector or cosine-corrected gravitational force, ΣF_c, between unit mass at the edge of a disk and the mass within the disk for every selected disk with radiuses increasing in steps of 5 units. Force increases up to the edge of the last disk at a distance of 150 units. Averaging and edge errors are larger on the first values but decrease rapidly as the cylinder flattens to a disk. This curve is computed, as shown above, for every point-like mass of the disk with Newton's equation 6.19 and cosine corrections. For example, a disk with R = 150 could have R_n values from 0 to 300 and θ_n divisions from 0 to 100 for a total number of 30,000 computation points on the disk. Note that Newton's gravitational equation, which applies directly only to point-like masses or homogeneous spheres, is valid in this case since the disk is divided in point-like masses.

The curve with triangles represents the cosine-corrected force, ΣF_c, between the whole disk of 150 units and the unit mass moved across the disk for every selected radiuses increasing in steps of 5 units. Force increases from zero at the left and up to a maximum at the edge of the disk at a distance of 150 units, where it is the same as the last value of the curve with diamonds. Then, it decreases sharply away from the edge. This curve shows values lower than force F_n curve (squares) near disk center and crosses it before reaching a maximum.

The curve with squares, F_n, represents the flat curve from equivalent center of mass force with a constant force value π up to the edge of the last disk at 150 units and then declining steadily away from the last disk. It is computed with Newton's equation 6.19, where disk mass increases as the square of the radius and force decreases as the square of the radius, thus yielding a constant value. But it is based on the false hypothesis that the mass of the disk is homogeneous like that of a sphere in 3-D. Contrary to the sphere, where Newton's equation can be solved with success, assuming that the whole mass of the sphere is at its center, this is not the case with a flat disk. Note that gravitational forces F_c and F_n are both 0, as shown on the graph, at the center of any homogeneous disk because of the symmetric cancellation of all the vectors.

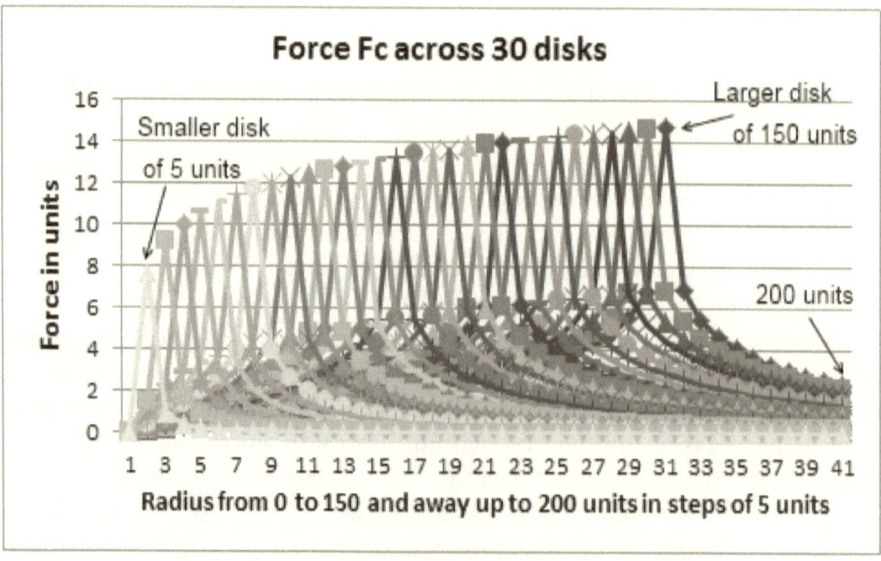

Figure 6.9 Gravitational force F_c across thirty different disks with radiuses from 0 to 150 units and away up to 200 units in steps of 5 units

Horizontal and vertical scales are the same as on figure 6.8. Last curve to the right on figure 6.9 (diamonds) is the same as the curve with triangles on the previous figure. Maxima of these thirty curves correspond exactly to the curve with diamonds from disk edges on figure 6.8. Each curve starts with zero at the center of a disk, reaches a maximum, and decreases toward zero away to the right. These *n* disks, with a density of one, may be combined with their *m* rings to form matrices from which forces, velocities, densities, and masses will be computed for real galaxies.

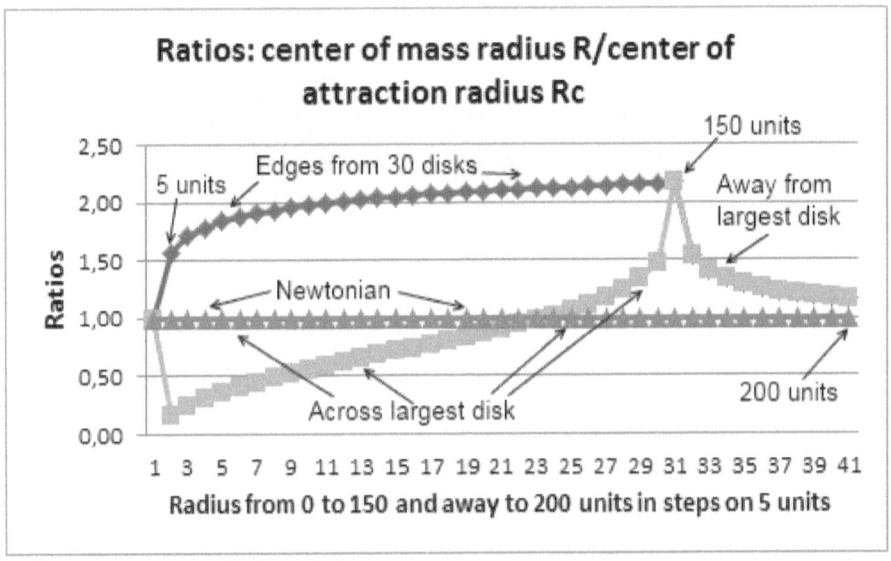

Figure 6.10 Ratios: center of mass radius over center-of-attraction radius, R/R_c, at the edge of thirty disks (diamonds), across the disk with radius R = 150 (squares), and Newtonian (triangles)

Horizontal scale is the same as on figure 6.9, and vertical scale represents ratios of center of mass radius over center-of-attraction radius, R/R_c.

The curve with diamonds ranges, on figure 6.10, from 1.0 at the center of the disk on the left to a maximum of about 2.2 at a distance of R = 150 on the right. The curve with squares also starts from 1.0 and varies, across the larger disk with R = 150 units, from a minimum of about 0.2 to a common maximum near 2.2, and then it swings down toward 1.0 away from the edge of the disk. Newtonian ratios (triangles) for the thirty disks stay flat at 1.0 all the way.

A value of 1 indicates that the center of mass and the center of attraction coincide, and a value of 2.0 indicates that the center of attraction is displaced from the center of mass to a location halfway between the test mass and the center of the disk. A lower value than 1 indicates that the center of attraction is displaced to the opposite side of the center of mass. Such displacements have consequences on velocity and mass computations.

Center of attraction at a distance R_c is computed with Newton's gravity equation 6.19, where both the mass of the disk and the cosine-corrected force of attraction, ΣF_c (top curve on figure 6.8), are known,

$$R_c = (MG/\Sigma F_c)^{\frac{1}{2}} \qquad (6.22),$$

where R_c is the radial distance from the test mass. Constant G is neglected on the curves.

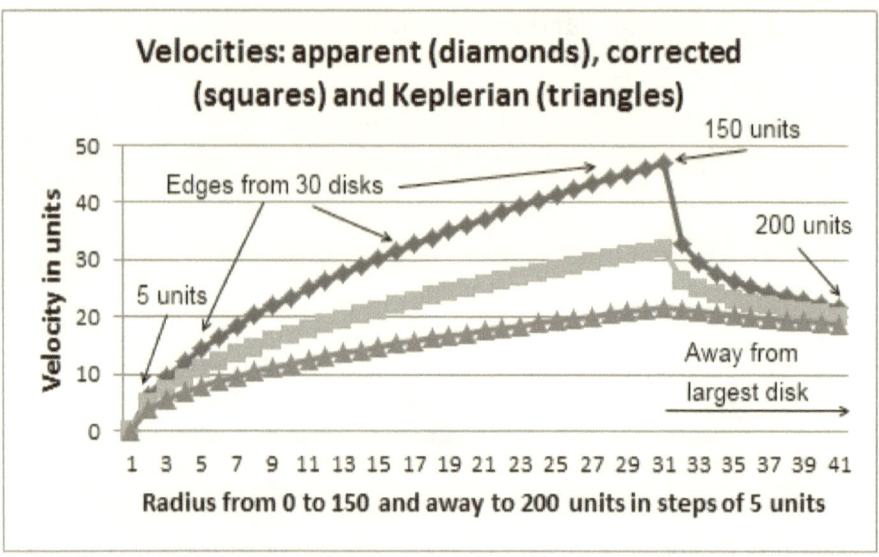

Figure 6.11 Top curve is apparent velocity, V_a; corrected velocity, V_c, is the middle curve; and Keplerian velocity, V_K, is the bottom one.

Horizontal scale is as above, and vertical scale represents velocity units. Each of the three velocities, V_a, V_c, and V_K, on figure 6.11, increases from the center at left and across the thirty disk edges, up to the largest at R = 150 on the right, and then drops down toward converging values away from the last disk.

The top curve is the apparent velocity, V_a, and it is the largest, about double the Keplerian velocity in this example, and would correspond in real life to best fit Doppler shift velocity evaluation across a galactic disk. Apparent velocity, V_a, is calculated from centripetal force equation

$$V_a = (F_c R/m)^{\frac{1}{2}} \qquad (6.23),$$

where the test mass is m = 1, F_c is the cosine-corrected gravitational force, and R is the radius. It is the apparent velocity at which the orbiting mass is moving around the center of mass. Corrected force, F_c, is used and not Newton's force F_n. (See figure 6.8.)

The middle curve represents the cosine-corrected velocity, V_c, about halfway between the other two. This is the corrected velocity at which the orbiting mass is moving around the center of gravitational attraction located at R_c. The latter is located closer to the orbiting mass than the center of mass. This corrected velocity is computed with the circular velocity equation

$$V_c = (GM/R_c)^{\frac{1}{2}} \qquad (6.24),$$

where G is the gravitational constant that is neglected, M is the mass inside the volume of the disk, and R_c is the corrected distance between the orbiting mass and the center of attraction, which varies according to the ratios on figure 6.10. This velocity will yield the corrected or real mass when combined with the corrected distance.

Finally, the Keplerian velocity, V_K, will also give the real mass inside a disk since it is based on the real total mass and total radius of the disk. The circular velocity equation 6.24 is also used to compute this velocity,

$$V_K = (GM/R)^{\frac{1}{2}} \qquad (6.25),$$

where G and M are the same as for the previous velocity, and R is the distance between the orbiting mass and the center of mass, which is also the geometrical center of the disk. The test mass is orbiting at this velocity about the center of mass. V_K is the velocity needed in an equivalent spherical Keplerian dynamic system to find correct mass values.

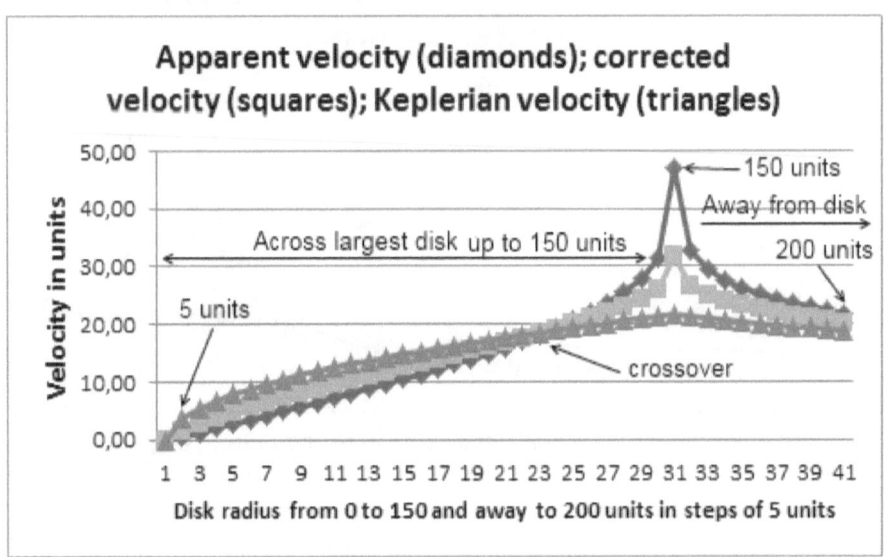

Figure 6.12 Comparison of apparent velocity, V_a (diamonds); corrected velocity, V_c (squares); and Keplerian velocity, V_K (triangles), for single disk of 150 units

Horizontal scale is as above, and vertical scale represents velocity units. Each of the three velocities, V_a, V_c, and V_K, on figure 6.12, increases from the center of this constant density disk at left, to the edge at R = 150 on the right, and then drops down toward converging values away from the edge. The maxima of these curves are the same values as for the disk with R = 150 on figure 6.11.

Diamonds represent apparent velocity, V_a, that would correspond in real life to best fit Doppler shift velocity evaluations across a homogeneous galactic disk. Since this curve crosses the other two curves, it means that velocities are undervalued near the center of the disk and overestimated toward the maximum.

Squares represent the cosine-corrected velocity, V_c, at which the orbiting mass is moving around the center of attraction. This velocity will yield the corrected mass when combined with the corrected distance.

Finally, the Keplerian velocity V_K (triangles) will also give the real corrected mass inside a disk since it is based on the real total mass and total radius of the disk. V_K is larger than V_a and V_c near the disk's center but crosses over and remains lower near the maximum and away from the edge.

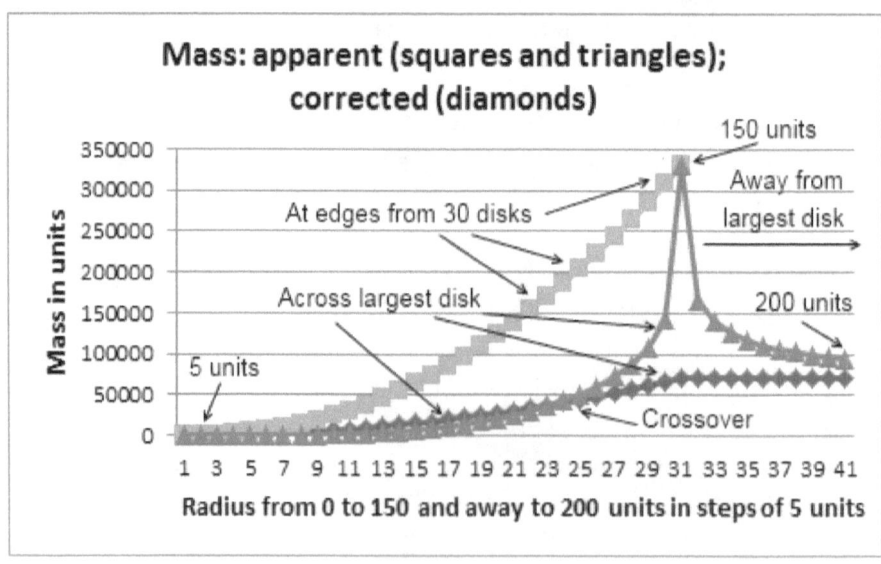

Figure 6.13 Corrected mass M (M_c or M_K, diamonds), apparent mass (squares) at the edge of thirty disks, and apparent mass (triangles) across largest disk

The horizontal scales of figure 6.13 are the same, and the vertical scale has a larger range than that on figure 6.5, where the calibrated mass was computed.

The curve with diamonds increases from 0 to 150 units and then is flat away from the edge of the last disk. It represents the cumulative calculated mass M, which is equal to corrected mass, M_c, or Keplerian mass, M_K, from velocity curves on figures 6.11. The circular velocity equation 6.24 may be used to compute this curve with $M_c = V_c^2 R_c/G$, where M_c is the corrected mass, V_c is the cosine-corrected velocity, and R_c is the corrected distance, or with equation 6.25, $M_K = V_K^2 R/G$, where M_K is the Keplerian mass; V_K, the Keplerian velocity; and R, the radial distance. Note that symbol M will be used in both cases since $M = M_c = M_K$. This mass curve has identical values as the calibrated mass from the original disk model on figure 6.5, thus confirming this method of computation whereby corrected mass values correspond to the real mass of a disk.

Apparent mass M_a (squares) at the edge of thirty different disks increases much faster than corrected mass M. It is computed with the circular velocity equation 6.24,

$$M_a = V_a^2 R/G \qquad (6.26),$$

where R is the radial distance and V_a is the apparent velocity, measured from Doppler shifts in real life. Apparent mass values are about 4.7 times larger than corrected mass at a radius of 150. This implies important errors of interpretation on the mass of galaxies when apparent mass is used instead of corrected mass.

The maximum value of the latter curve is the same as on the curve with triangles, which represents apparent mass, M_a, calculated from corrected force, F_c, and apparent velocity, V_a, across the larger disk with R = 150 and up to fifty units away from the edge. This curve crosses the curve of the corrected mass M, indicating that apparent mass would yield a mass deficit toward the center of the disk and a large mass excess toward the edge of the same disk. It shows a sharp break to lower values at the edge of the disk but remains larger than mass M while converging. Obviously, apparent mass is not real since it shows an impossible mass decrease away from the edge of the disk. Mass is positive and additive within a sphere or a disk and can only increase away from the center of a body or set of bodies. Negative mass was never recorded.

In a nutshell, the loops from mass M at the start to mass M at the end of this theoretical example may be summarized as follows on figure 6.14:

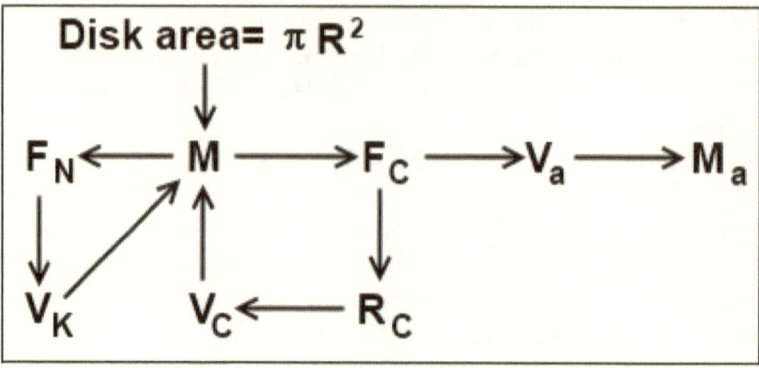

Figure 6.14 Summary of the theoretical model

From the radius of a disk, mass is found to compute the corrected force from which apparent velocity and mass are calculated and also the corrected radial distance and velocity that will yield the original mass value. The original mass also gives the Newtonian force and then the Keplerian velocity to find again the original mass. This completes the proof on the calibrated mass for the thirty selected disks with radius R = 0 to 150 units and the single selected disk with radius R = 150 units and k = 5 to 50 units away from the edge of the disk, computed in forty-one steps of 5 units, all with a thickness of 1 unit and a density of 1.

These results may be generalized and applied to any flat disk. They may also be combined with many disk layers of various sizes and/or the presence of a central bar, core, bulge, and more or less curved spiral arms in a stack of disks to simulate all kinds of galaxies.

6.1.5 Mass, Velocity, and Distance Ratios

Combining mass, velocity, and distance values, it is possible to establish a few useful ratios applicable to our disk models as well as to all galactic disk computations.

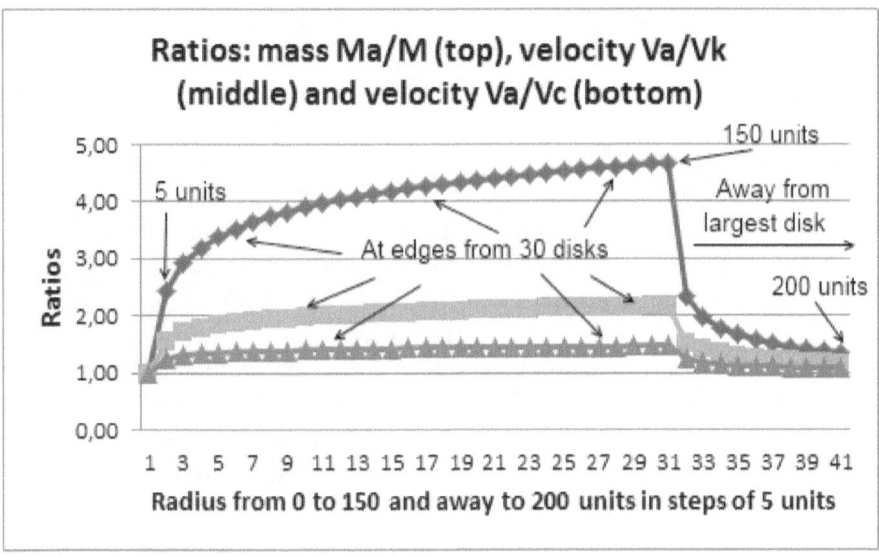

Figure 6.15 Mass ratios, M_a/M, at the top; velocity ratios, V_a/V_K, or distance ratios, R/R_c, in the middle; and velocity ratios V_a/V_c at the bottom

Horizontal scale is the same as on the previous figure. Vertical scale shows ratios. All three ratio curves shown on figure 6.15 are based on disk models with density 1 and radiuses between R = 0 and 150 units computed in steps of 5 units and with 50 units away from the edge of the largest disk.

The bottom curve represents apparent-to-corrected-velocity ratios V_a/V_c at the edge of thirty separate disks. It reaches a maximum near a value of 1.4 at the edge of the largest disk and then decreases.

The middle curve with a maximum value close to 2.2 corresponds to two identical curves showing the same ratios for apparent-to-Keplerian velocity V_a/V_K and also center of mass to center of attraction distance R/R_c already presented on figure 6.10.

Values of the top curve are the apparent-to-Keplerian mass ratios M_a/M. They reach a maximum value of about 4.7. The circular velocity equations (6.25 and 6.26) give $M = V_K^2R/G$ and $M_a = V_a^2R/G$, from which apparent mass-to-corrected-mass ratios M_a/M are found equal to the square of apparent-to-Keplerian velocity ratios V_a^2/V_K^2:

$$M_a/M = V_a^2/V_K^2 \qquad (6.27).$$

We take Newton's equation, $R = (GM_a/F_c)^{\frac{1}{2}}$, and equation 6.22, where $R_c = (GM/F_c)^{\frac{1}{2}}$, that we combine with the circular velocity equations 6.25 and 6.26 as above to yield a very useful result: apparent-to-Keplerian velocity ratio $V_a/$

V_K is equal to the ratio between the center of mass and the center of attraction, R/R_c,

$$V_a/V_K = R/R_c \qquad (6.28),$$

from which center of attraction radius, R_c, can be determined when apparent (Doppler shift equivalent) velocity, V_a; Keplerian velocity, V_K; and radius R are measured values from, for example, galactic studies and descriptions.

Then, the centripetal force equation 6.23 and $V_c = (F_c R_c/m)^{1/2}$ yield the centripetal acceleration per mass unit:

$$a = V_c^2/R_c = V_a^2/R \qquad (6.29).$$

But for the equivalent spherical model with a Newtonian force, F_n, and a Keplerian velocity, V_K, we would find the Newtonian acceleration, $a_n = V_K^2/R$.

Also, combining equations 6.27 and 6.28, we find the following useful result: the square of center of mass over center of attraction ratios R^2/R_c^2 is equal to apparent mass-to-corrected-mass ratios M_a/M,

$$R^2/R_c^2 = M_a/M \qquad (6.30),$$

from which corrected mass M may be calculated when R^2, R_c^2, and M_a are known variables.

Finally, combining equation 6.28 with equation 6.29, we find a special velocity relationship: corrected velocity, V_c, is equal to the square root of the product of apparent velocity, V_a, with Keplerian velocity, V_K,

$$V_c = (V_a V_K)^{1/2} \qquad (6.31),$$

from which corrected velocity, V_c, can be found since V_a is calculated from Doppler shifts in galactic evaluations, and V_K may be computed from a spherical model of mass/luminosity distribution.

6.2   Application to Four Typical Galaxies

In order to apply the results of the theoretical model to real-life galaxies, a more appropriate model must be developed to imitate the gradually increasing mass or density of a galactic disk from the edge to the center.

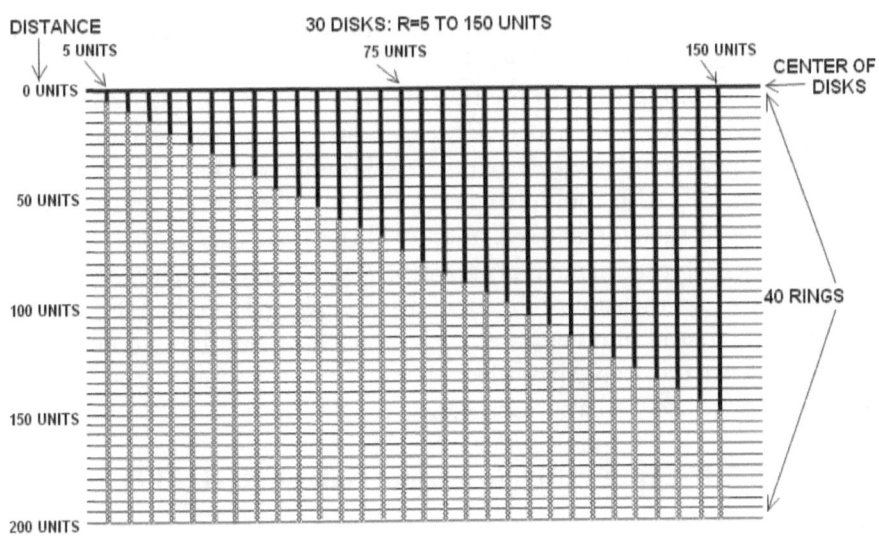

Figure 6.16 Matrix set up to compute galactic forces, velocities, densities, and masses, using thirty disks with R = 0 to 150 units in steps of 5 units

The basic *m*-by-*n* matrix set-up used to compute forces, densities, masses, and velocities for spiral galaxies is shown on figure 6.16 with n = 30 disks represented by black vertical lines with radiuses varying from 0 to 150 units in steps of 5 units. Three of these disks are specifically indicated as 5, 75, and 150 units. Disks correspond to the presence of masses starting at the center (top horizontal line) and extending down to their respective edges, maximum 150 distance units. Gravitational forces are present away from these disks and are represented by gray line extensions away from the edges up to a maximum of 200 units in steps of 5 units. Horizontal lines correspond to m = 40 rings, including 30 rings for the largest disk up to 150 units and 10 rings for its extension up to 200 distance units. When a value of zero is included at the center, a total of forty-one rows is involved.

Three basic matrices, *A*, *B*, and *C*, are combined to compute the stack of the total force per ring, total mass, and total density for the disks. Matrix *A* (forty-one rows by thirty columns) contains gravitational force coefficients F_c for each ring element of each disk with a density of one and their extensions away from the edges. Matrix *B* (forty rows by thirty columns) contains density coefficients ρ found iteratively by matching apparent velocities from Doppler shifts with computed apparent velocities using equation 6.23, $V_a = (\Sigma F_c R/m)^{\frac{1}{2}}$. Matrix *C* (thirty-one rows by thirty columns) contains basic mass coefficients for each ring element of each disk.

Newton's equation says that force, F, is proportional to mass, M, implying that any increase of mass by a density factor will increase the value of force, F, in the same proportion, so multiplying each element of matrix A by the corresponding element of matrix B yields matrix D, which represents total force for each element. These elements are stacked horizontally to get the sum of total forces, ΣF_c. Also, multiplying each element of matrix B (density) by the corresponding element of matrix C (mass) will yield matrix H, which represents total mass for each ring element of each disk. Again, these elements are stacked horizontally to get the sum of the total mass in each galactic ring. Finally, the mass of each ring is summed to obtain corrected or real galactic mass from which corrected and Keplerian velocities are computed simultaneously in the same iteration process on a spreadsheet.

6.2.1   Application to a Large Galaxy: NGC 3198

Galaxy NGC 3198 is a high-surface-brightness galaxy located in the Ursa Major constellation, about 14.4 Mpc away, with a radius of about 29.7 kpc and classified as a spiral with a small core, an Sc galaxy.

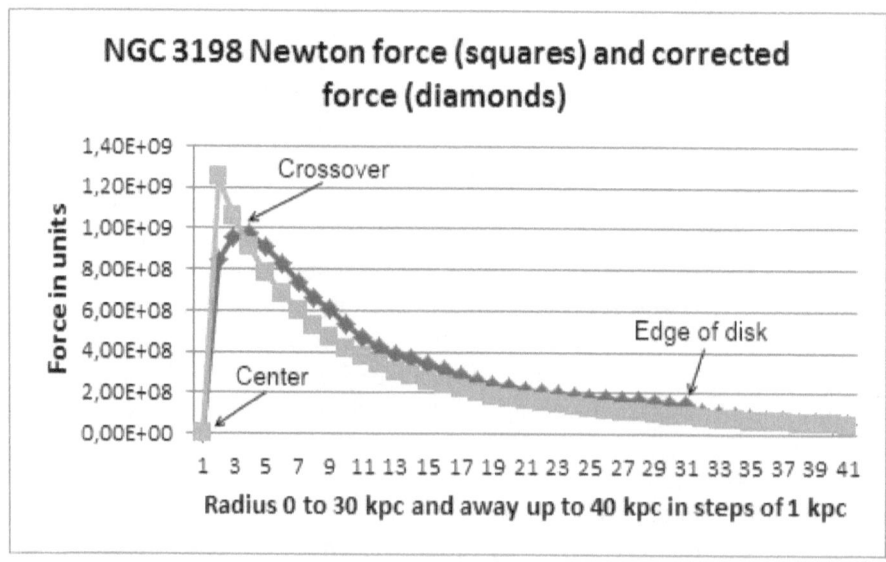

Figure 6.17 Newton force and corrected force for NGC 3198

Galactic radius for NGC 3198 runs, from left to right on the horizontal scale, from 0 to 30 kpc, with an extension of 10 kpc away from the edge of the disk. Note that horizontal distance scale is converted from the previous theoretical model with a factor of five units for 1 kpc. The vertical scale represents force in units on figure 6.17.

The distribution of Newton force, F_n, is shown with the square symbols: it is computed with Newton's universal law of gravitation and assumes that the mass across the disk is concentrated at the center and thus decreases with the square of the radius. On the other hand, the diamond symbols represent the corrected force, F_c, which is computed from the summation of the force distribution in a stack of thirty disks. Except near the center, where an important mass must be present in the bulge and must give some similitude to the curve from Newton's force, it generally shows a larger and gradual decrease of force F_c across the galaxy, corresponding to a gradual decrease of mass across the disk. We find $F_n = F_c$ at the center and at the crossover, $F_n > F_c$ between the center and the crossover, and $F_n < F_c$ between the crossover and the edge of the disk and even away from the disk.

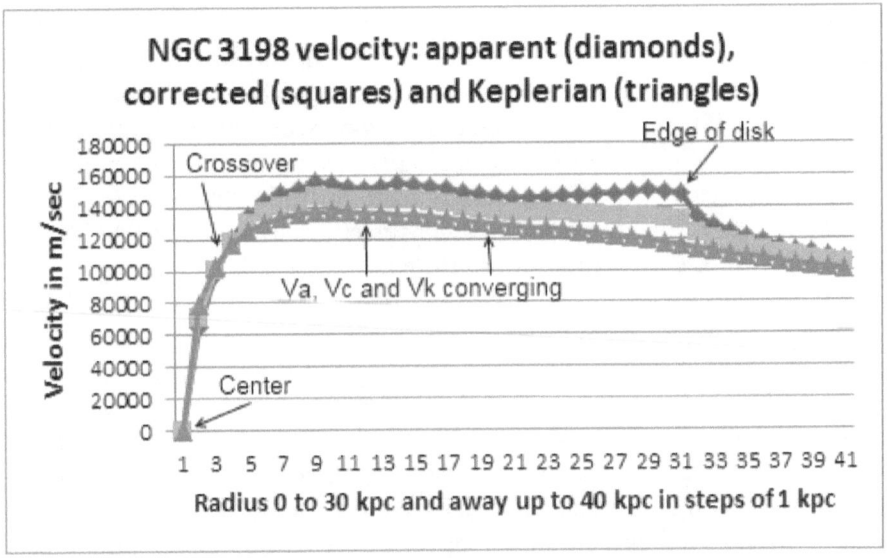

Figure 6.18 NGC 3198 apparent velocity, V_a (diamonds); corrected velocity, V_c (squares); and Keplerian velocity, V_K (triangles), from a stack of thirty disks

The horizontal scale on figure 6.18 is the same as on the previous figure, and the vertical scale shows velocity in m/s.

The curve with diamonds from 0 to 30 kpc represents rotation or apparent velocities as extracted from Doppler shift measurements. An average is presented here from one of the most published velocity curve (Begeman 1989)

and plotted against distance from the center in kpc. The last ten values are computed from equation 6.23.

The curve with squares corresponds to corrected velocity, V_c. It may be computed from the circular velocity equation 6.24 with $V_c = (GM/R_c)^{1/2}$, where M is the corrected mass and R_c is the center-of-attraction radius. The last ten values are computed away from the edge of the galaxy. These velocity values, as for the theoretical model, yield corrected mass for a disk when combined to center-of-attraction values. These are the velocities at which bodies are orbiting around their own center of attraction across the galactic disk.

The Keplerian velocity curve (triangles) represents the equivalent Keplerian spherical dynamics for NGC 3198 as if the center of mass were the center of attraction. It is calculated from cumulative matrix H corrected mass values. The last ten Keplerian velocity values are also calculated away from the edge. When reduced to a spherical model, masses seem to be orbiting with velocity V_K around the center of mass although they really orbit according to V_c around R_c, the ΔV vector difference being proportional to the torque down the spiral curve toward the center. This velocity curve is generally lower than the other two except for the three first values near the center of the galaxy, which implies that apparent (Doppler shifts) velocities are too small near the center and too large toward the edge with corresponding consequences on mass evaluation. Note that this curve is gradually decreasing to the right after a maximum and is not a flat curve like that from apparent velocity, thus indicating the absence of mass excess or dark matter toward the edge of the disk.

When velocities V_a and V_c converge, as observed at about ¼ and ½ of the curves, resulting forces bring masses closer to a circular pattern and may produce a density increment or the equivalent of a traffic jam, where new stars may be produced. But when velocities are equal, such as observed at the crossover on figure 6.18, masses are locked in and moving about a center of mass, which coincides with the center of attraction, so linear and tidal forces may have more influence and a bar may start to form. Finally, when V_c is larger than V_a as observed near disk's center, then the mass associated with V_c may tend to diverge from the center.

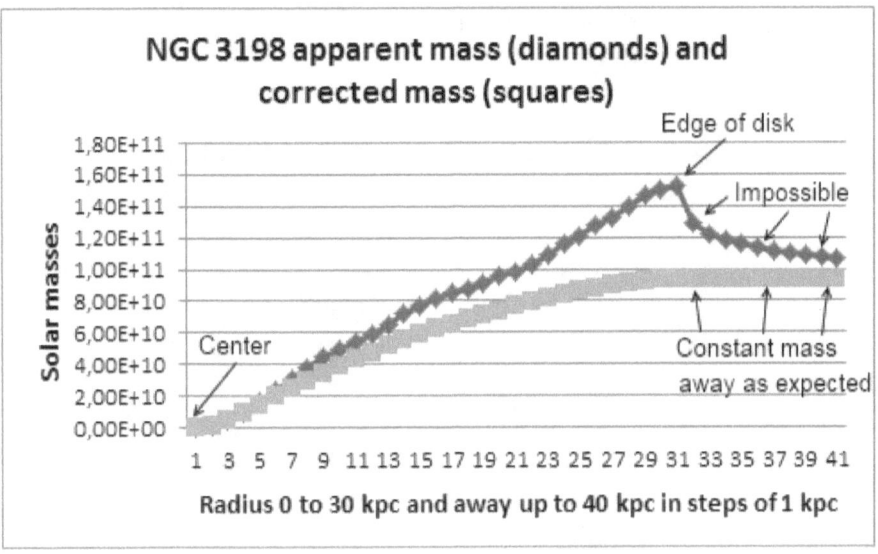

Figure 6.19 Apparent mass, M_a, on the top and corrected mass, M, at the bottom for NGC 3198, using a thirty-disk model

The horizontal scale of figure 6.19 is the same as on the previous figure, and the vertical scale represents mass in solar masses (sun = 1.99×10^{30} kg).

The top curve represents apparent mass, M_a, which is lower than the corrected mass for the first three values but increases generally faster than the corrected mass in this real example. It shows a sharp break to lower values at the edge of the disk but remains larger than the corrected mass while converging. Apparent mass values reach a maximum of 150 billion solar masses and thus are 1.60 times larger than the corrected mass with a maximum of 94 billion solar masses at a radius of 30 kpc. Apparent mass is computed with the circular velocity equation 6.26, $M_a = V_a^2 R/G$, where G is the gravitational constant, R is the radial distance, and V_a is the apparent velocity, measured from Doppler shifts or from matrix D. Apparent mass is not correct since it shows an impossible mass decrease away from the edge of the disk. As mentioned above, mass is positive and additive within a sphere or a disk and can only increase away from the center of a body or set of bodies.

The bottom curve increases from 0 to 30 kpc and then, since no mass is added outside the visible disk, is flat away from the disk over the next 10 kpc. It represents the cumulative calculated mass M from velocity curves V_c or V_K on figure 6.18 and, respectively, R_c or R distance. This mass curve corresponds to corrected or real mass for galaxy NGC 3198. The circular velocity equation 6.24 or an equivalent may be used to compute the bottom curve M with M =

$V_c^2 R_c/G$, where G is the gravitational constant, M is the corrected mass, V_c is corrected velocity, and R_c is corrected distance.

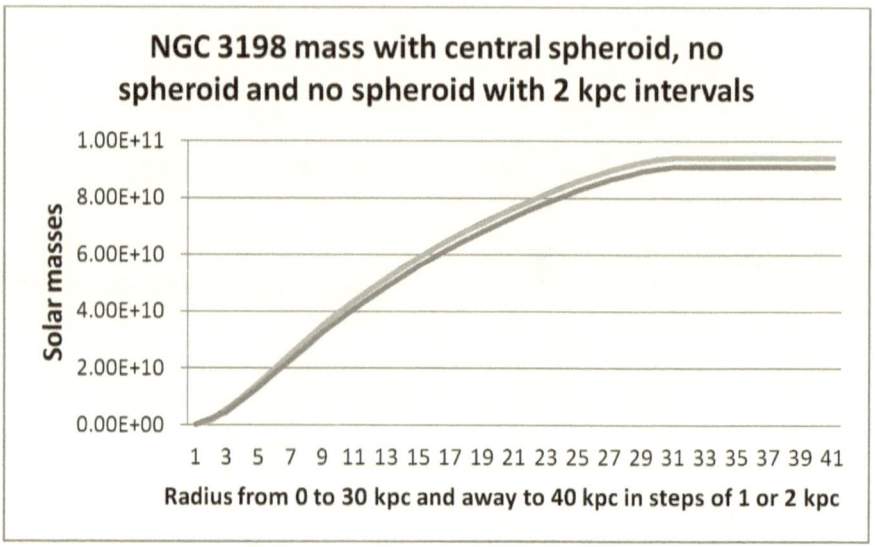

Figure 6.20 Three curves from NGC 3198 corrected or real mass with central spheroid, no spheroid, and no spheroid with 2 kpc intervals

Three different thin curves of the corrected or real mass are presented on figure 6.20 to test the sensitivity of the model to the presence or not of a central spheroid and to computation intervals of 2 kpc instead of 1 kpc with no spheroid. Central spheroid model (top curve) is calculated every 1 kpc with a stack of twenty-nine disks plus the first 1 kpc evaluated with classical Kepler-Newton's spherical model to account for the bulge at the center. The curve without a spheroid, almost indistinguishable from the first one, is computed every 1 kpc, exclusively with the complete stack of thirty disks. The third curve, generally located below the other two curves, represents rougher values computed every 2 kpc with a stack of fifteen disks and without a central spheroid. Total mass difference computed in these three ways amount to a maximum variation of about 3% for the 2 kpc intervals at the edge of the galactic disk. NGC 3198 corrected mass is thus evaluated at 94 billion solar masses.

Two papers published by Gallo and Feng [84] on galactic rotation with a thin-disk model similar to the above give a very close value of 99 billion solar masses for NGC 3198 and 140 billion solar masses for the Milky Way, also very close to our value of 132 billion solar masses. They used integrals to accomplish the summation in their model and found concordant results, including the conclusion that dark matter is repelled and not necessary to explain the galactic dynamics.

The same exercise will be applied to the small galaxy NGC 4062 to discover its force profile, velocity, and mass distribution.

6.2.2   Example from a Small Galaxy: NGC 4062

William Herschel discovered this small galaxy in 1787 in the Ursa Major constellation. It is classified as a small-nucleus Sc galaxy or as a small barred SBc galaxy with a radius of about 4 kpc and has a flocculent appearance.

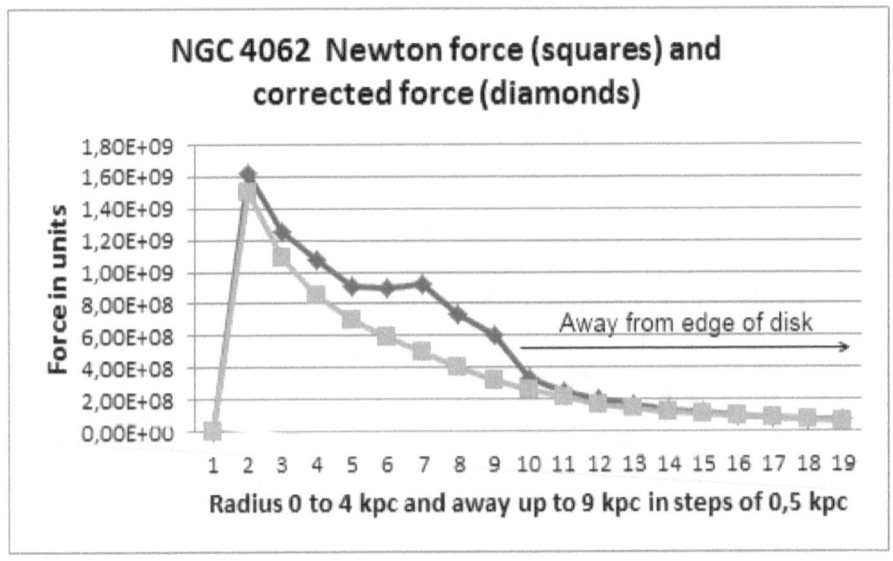

Figure 6.21 Newton and corrected force for NGC 4062

No crossover is observed between Newton's and corrected force values, shown respectively as squares and diamonds on figure 6.21. The large hump associated with the corrected force is related to the presence of a large mass within the disk at a distance between about ½ to ¾ from the center and decreasing up to the edge. This may indicate the recent merging of some neighboring material that is in the process of integration to the galaxy.

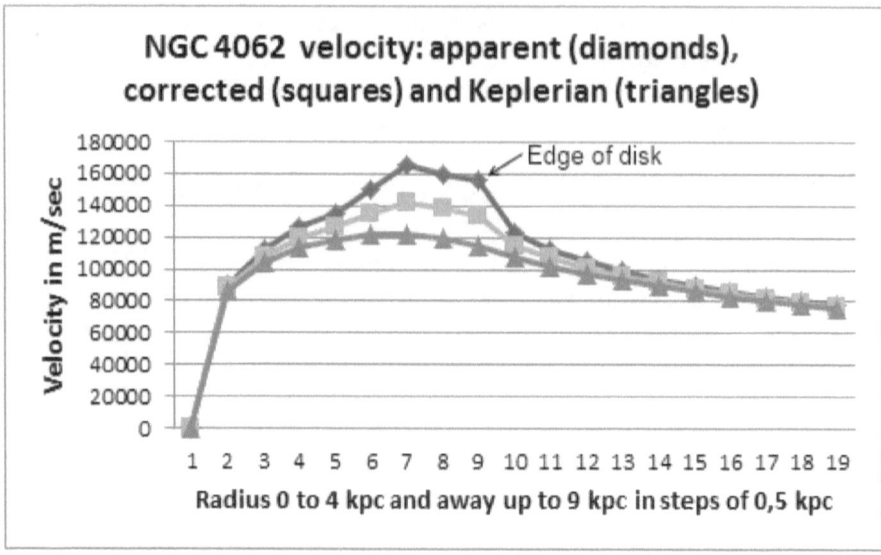

Figure 6.22 Velocity distributions for NGC 4062

The velocity distributions of NGC 4062 on figure 6.22 do not show a crossover and are concordant with force distributions. Apparent velocities (diamond symbols) indicate a sharp increase to 166 km/s between ½ and ¾ of the radius, followed by a decrease to the edge of the disk at about 156 km/s. The same pattern is observed with the corrected velocity (squares), but with a maximum of 142 km/s and less pronounced variations. Finally, the Keplerian velocity (triangles) with a maximum of 122 km/s always shows the smoothest curve with the lowest values since it is based on the Newtonian concept that the whole mass is like a point at the geometric center.

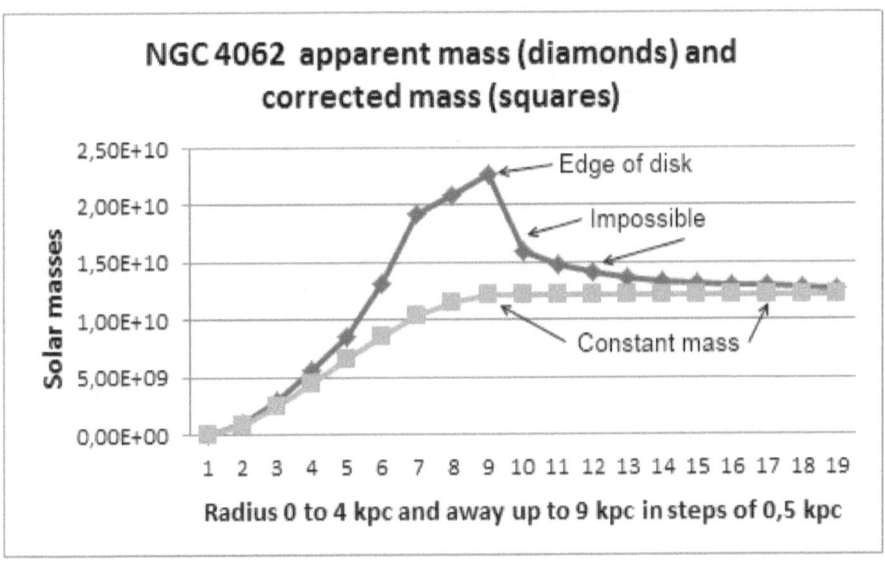

Figure 6.23 Apparent and corrected mass for NGC 4062

The apparent and corrected mass distribution is plotted on figure 6.23. As expected from force and apparent velocity variations, we find a sharp increase in the apparent mass at ½ to ¾ of the radius that is also present in the corrected mass. Apparent mass reaches a maximum of 22.7 billion solar masses, but the decrease away from the edge of the disk is impossible and indicates that apparent mass is incorrect. Corrected mass shows a constant mass value of 12.2 billion solar masses at the edge and away from the galaxy. Thus, apparent mass is an impressive 1.86 times more than corrected mass. This is a large error in mass evaluation, especially if we could also take into account the presence of poorly visible low-surface-brightness satellites and low-resolution neighboring clouds not accessible in the range of frequencies most commonly used in astronomy.

6.2.3   Application to a Barred Galaxy: NGC 4389

Galaxy NGC 4389 is located in the constellation Canes Venatici and has an apparent magnitude between 11.7 and 13. It was discovered by William Herschel in 1788 and is classified as a small to intermediate barred spiral and SBbc galaxy. We include this galaxy as a model for a typical barred galaxy although a special treatment may be necessary due to the fact that a bar is not a spheroid or a disk but an elongated structure with some specific characteristics.

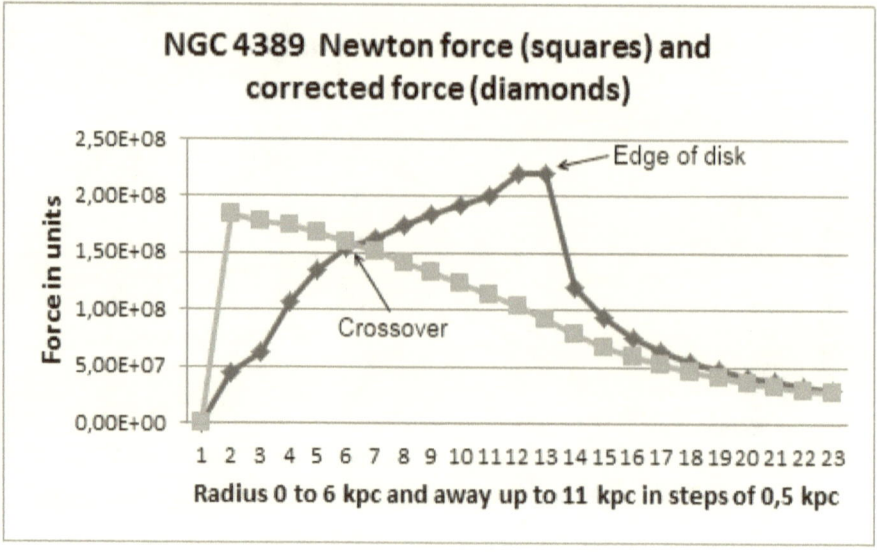

Figure 6.24 Newton and corrected force for NGC 4389

A strong crossover, located at a distance of 2.5 kpc (about 40%) from the center, is observed between Newton's force (squares) and corrected force (diamonds) on figure 6.24. The crossover is located near the edge of the bar. The slope of the corrected force is very steep between the center and the crossover indicating the presence of a bar. It rises with a lower slope to the edge of the galaxy, which points to the presence of an important mass spread all the way to the edge at 6 kpc.

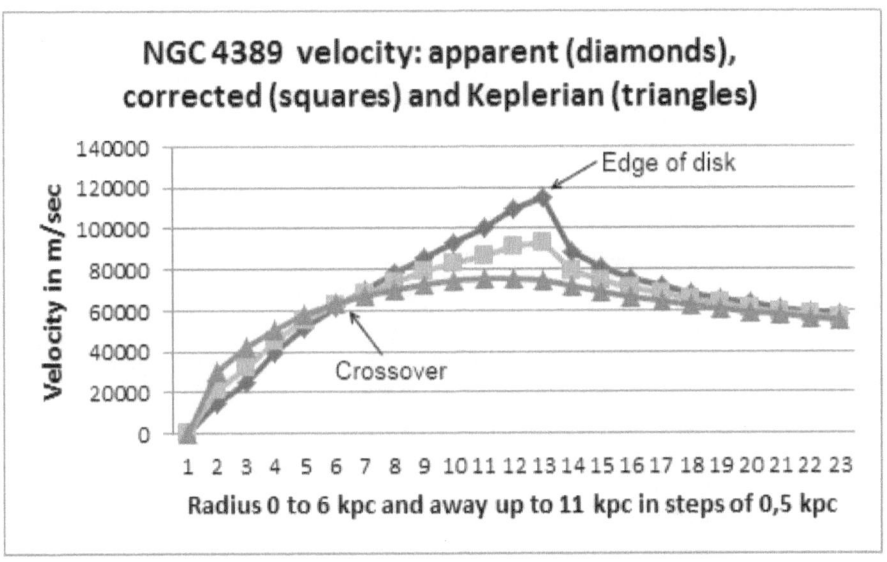

Figure 6.25 Velocity distributions for NGC 4389

The velocity distribution of NGC 4389 on figure 6.25 shows a crossover, which is concordant with force distributions. Apparent velocities (diamond symbols) indicate a two-step increase to the maximum at 115 km/s. A similar pattern is observed with the corrected velocity (squares), but with a maximum of only 93 km/s. Finally, the Keplerian velocity (triangles) with a maximum of 76 km/s shows a smooth curve since it is based on the Newtonian concept of the point mass at the center. The crossover represents a critical point where all three velocity curves have the same value. Between the center and the crossover, we find that apparent velocity may be undervalued since it is less than the corrected one, and from the crossover to the edge, apparent velocity would be overestimated. As expected, and to keep a constant angular velocity between the bulge and the crossover, both apparent and corrected velocities increase approximately as the radius across the bar.

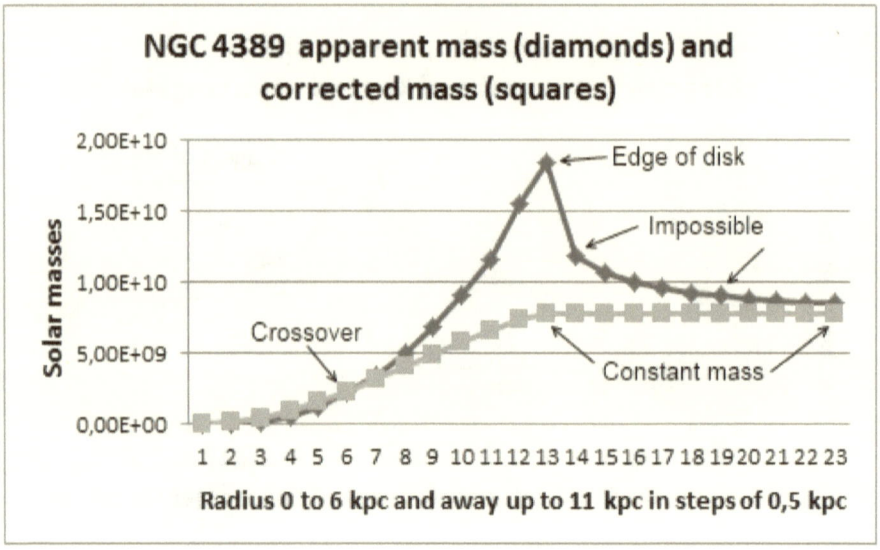

Figure 6.26 Apparent and corrected mass for NGC 4389

Figure 6.26 shows apparent and corrected mass distribution. As expected from force and apparent velocity variations, we find a sharp increase in the apparent mass between the crossover and the galactic edge while the corrected mass curve shows a more steady increase even before the crossover. Apparent mass reaches a maximum of 18.5 billion solar masses, but the decrease away from the edge of the disk is impossible and indicates that apparent mass is incorrect. Corrected mass shows a constant mass value of 7.75 billion solar masses at the edge and away from the galaxy. Thus, apparent mass is an impressive 2.39 times more than corrected mass. This is a major error in mass evaluation, especially when we consider the influence on the total mass of a galaxy from other factors, such as poorly visible galactic and gas material in the neighborhood that is not yet accessible with present-day technology.

6.2.4   Example from Elliptical Galaxy NGC 3379

Elliptical galaxy NGC 3379 was selected to verify how much a disk model could apply to an apparent spheroidal-shape galaxy. NGC 3379 was discovered in the Leo constellation by Pierre Méchain in 1781. It is classified as an elliptical galaxy E1, but since an arm or a circle of dusty material is seen on some published pictures, it could also be a face-on S0 galaxy. In the latter case, the disk model should partly apply.

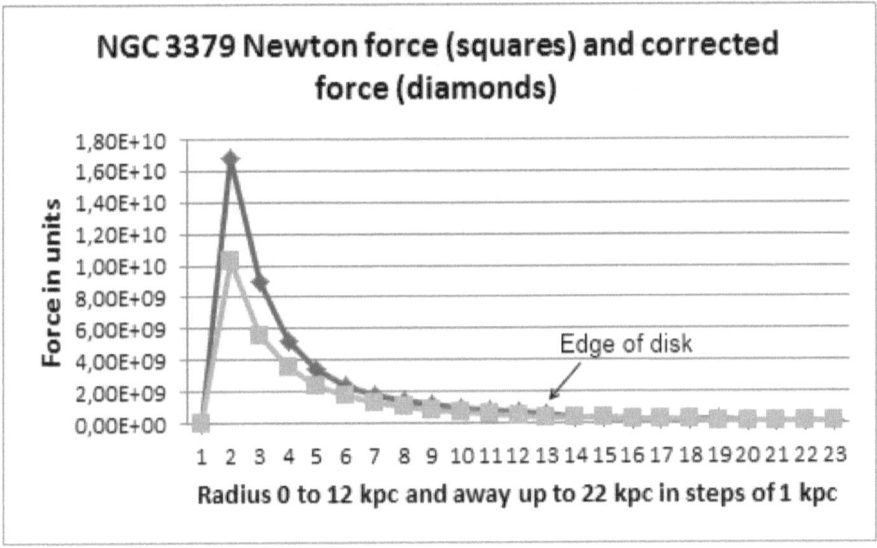

Figure 6.27 Newton and corrected force for NGC 3379

Both curves from Newton's force (squares) and from the corrected force (diamonds) follow a similar Newtonian decrease away from the center on figure 6.27, betraying the presence of an extremely large mass near the center of NGC 3379. No crossover is occurring, but values from the corrected force clearly dominate the distribution and indicate the presence of some flattening likely intermediate between disk and spheroid models.

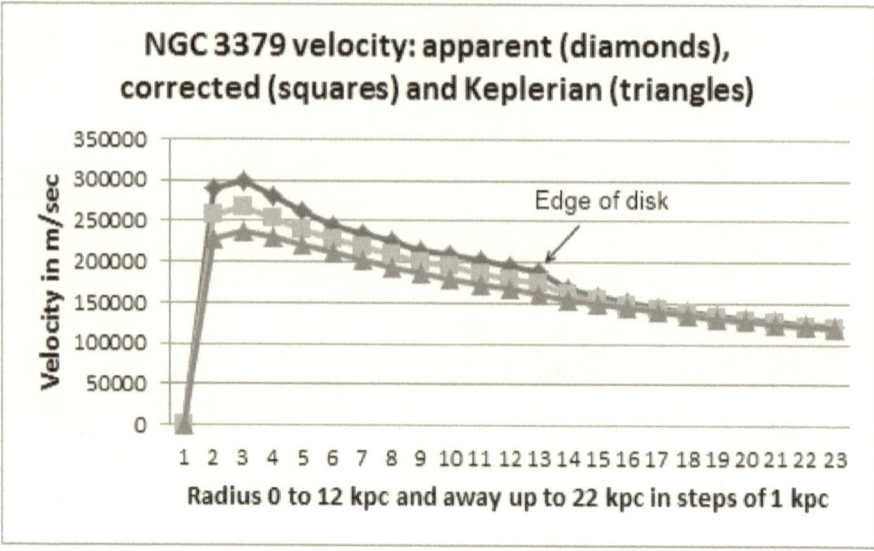

Figure 6.28 Velocity distributions for NGC 3379

No crossover is present in the velocity distribution of NGC 3379 on figure 6.28. Contrary to the typical small galaxy NGC 4062 and to the barred galaxy NGC 4389, which reached their maximum velocities toward the edge of the disk, we find a maximum close to the center for NGC 3379, which makes it similar to the large galaxy NGC 3198. But NGC 3379 shows a large apparent velocity excess near the center and tapering down toward the edge while NGC 3198 had the large excess close to the edge. Thus, we expect a smaller mass contrast between apparent and corrected velocities, less than for the large galaxy NGC 3198.

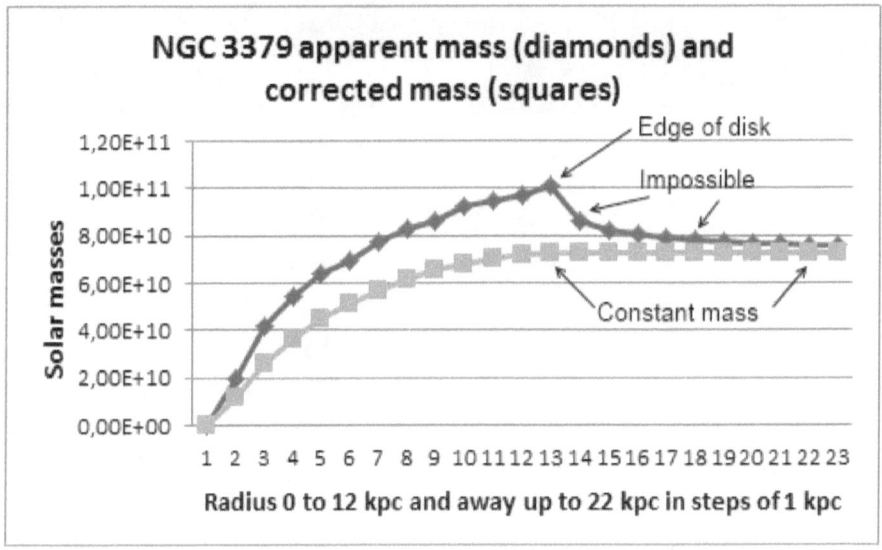

Figure 6.29 Apparent and corrected mass for NGC 3379

Figure 6.29 shows apparent and corrected mass distribution for NGC 3379. As expected from force and apparent velocity variations, we find a more rapid increase in the apparent mass near the center and much less added mass toward the edge while the corrected mass curve shows a more steady increase. Apparent mass reaches a maximum of 101 billion solar masses at 12 kpc, but the decrease away from the edge of the disk is also impossible and indicates that apparent mass is incorrect. Corrected mass shows a constant mass value of 72.8 billion solar masses at the edge and away from the galaxy. Thus, apparent mass is an important 1.39 times more than corrected mass. This is a remarkable error in mass evaluation, especially when we realize that we are dealing with an elliptical or possibly an S0 galaxy, where an intermediate model between disk and spheroid may apply.

6.3   Statistical Results with Forty-Six Galaxies

What would be the outcome if we apply our method and model on a larger number of galaxies [38], with some statistical significance? To answer this question, forty-six galaxies were selected from various published data, of which sixteen are low-surface-brightness galaxies (LSB) [32] and thirty are high-surface-brightness galaxies (HSB) [29] [47].

	HSB galaxies:			LSB galaxies:	
Name of galaxy	Galactic radius in kpc	Comments		Name of galaxy	Galactic radius in kpc
NGC 1003	1.0		DDO 64		2.5
NGC 4096	1.5		DDO 168		4.0
NGC 4448	2.0		UGC 1281		5.0
NGC 4062	4.0	Small selected	NGC 2366		5.5
NGC 2708	4.5		NGC 4455		6.0
NGC 3495	5.0		F 583-4		6.5
NGC 4389	6.0	Barred selected	UGC 731		7.0
NGC 3691	7.5		DDO 170		7.5
NGC 2259	8.0		NGC 100		8.0
NGC 3949	9.0		DDO 154		8.5
NGC 1417	10.0		F 568-3		11.0
NGC 3379	12.0	Elliptical selected	UGC 4173		12.0
NGC 4527	13.0		F 571-8		14.0
NGC 3917	15.0		UGC 711		16.0
NGC 2590	16.0		UGC 3137		26.0
NGC 4217	17.0		UGC 1230		36.0
NGC 2403	19.0				
NGC 3079	20.0				
NGC 3031	21.0				
NGC 4088	22.0				
NGC 3521	23.0				
NGC 2903	25.0				
NGC 6946	29.5				
NGC 3198	30.0	Large selected			
NGC 3992	32.0				
NGC 3726	34.0				
NGC 7331	36.0				
NGC 224	38.0				
NGC 5055	40.0				
NGC 2841	44.0				

Table 6.1 List of thirty HSB and sixteen LSB galaxies with radius in kpc

In column 1 of table 6.1, we find the list of thirty HSB galaxies with their respective NGC denomination, followed in column 2 by their maximum radius in kiloparsec (kpc) identified with a Doppler redshift signature that could be corrected and converted to apparent velocities as found in the literature. Column 3 indicates the four selected galaxies that were previously presented as typical galaxies. The list of LSB galaxies is found in column 4 with their respective maximum radius listed in column 5.

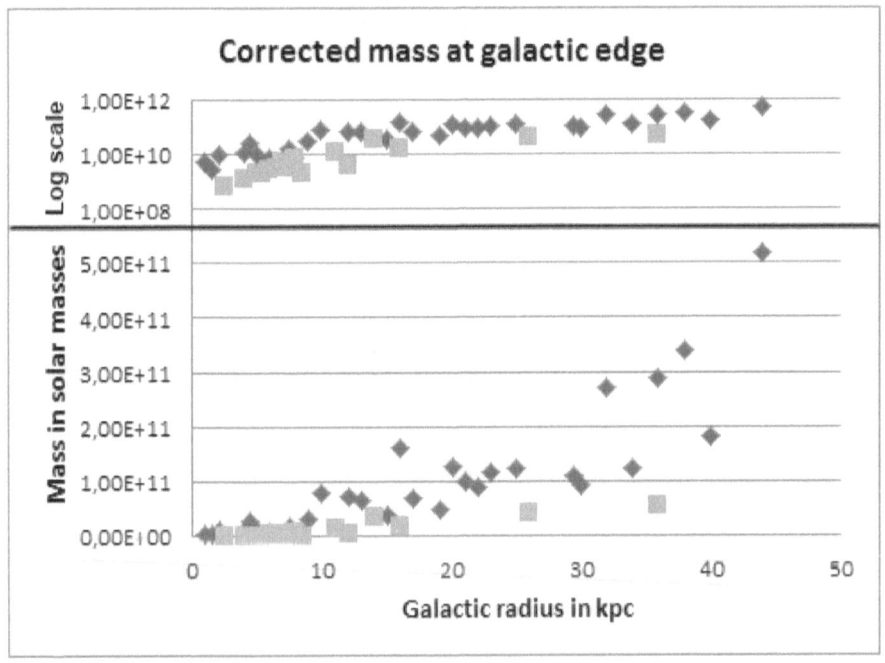

Figure 6.30 Corrected galactic mass at the edge of selected galaxies

The corrected galactic mass at the edge of forty-six selected galaxies is presented on figure 6.30 with one common horizontal scale and two different vertical scales. The horizontal scale is from 0 to 50 kpc, and the two vertical scales are in solar masses, but the top one is a logarithmic scale to show the better resolution for galaxies with a smaller radius. The black diamond symbols represent HSB galaxies while the gray squares represent LSB galaxies. We observe that corrected galactic mass increases with the radius of galaxies and would expect this increase to be in proportion to the square of the radius for a uniform disk. But the mass is increasing with less than the square of the radius since mass is not uniform in a galaxy and decreases generally and gradually from the massive bulge at the center. The most important result is that the mass of LSB galaxies is lower than the mass of HSB galaxies all the way across the distribution. Having less mass and, thus, less star-forming clouds and regions seems a perfect reason for a galaxy to exhibit low surface brightness.

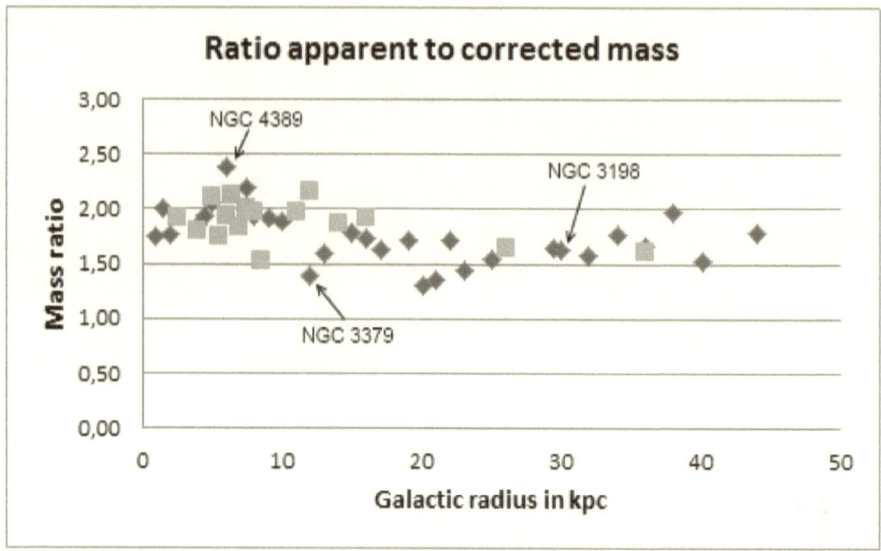

Figure 6.31 Ratio between apparent and corrected mass for forty-six galaxies

Ratios between apparent and corrected masses are presented on figure 6.31 for the forty-six selected galaxies. The vertical scale shows that all ratios are found between 1 and 2.5. Both HSB and LSB galaxies follow the same trend, with decreasing apparent-to-corrected-mass-ratios across the distribution from about 2 to 1.5. The curve indicates larger ratios for small galaxies and seems to be flattening at larger radiuses. The position of three galaxies discussed above is indicated as a reference. We note that NGC 4389, the barred galaxy, has the highest apparent-to-corrected-mass ratio at 2.39, perhaps indicating that large barred galaxies may contain the largest errors in mass estimates or the need to improve the model when a bar is present. On the other hand, the elliptic galaxy NGC 3379 shows a low apparent-to-corrected-mass ratio, perhaps indicating a drift from the disk model to a more spheroidal model.

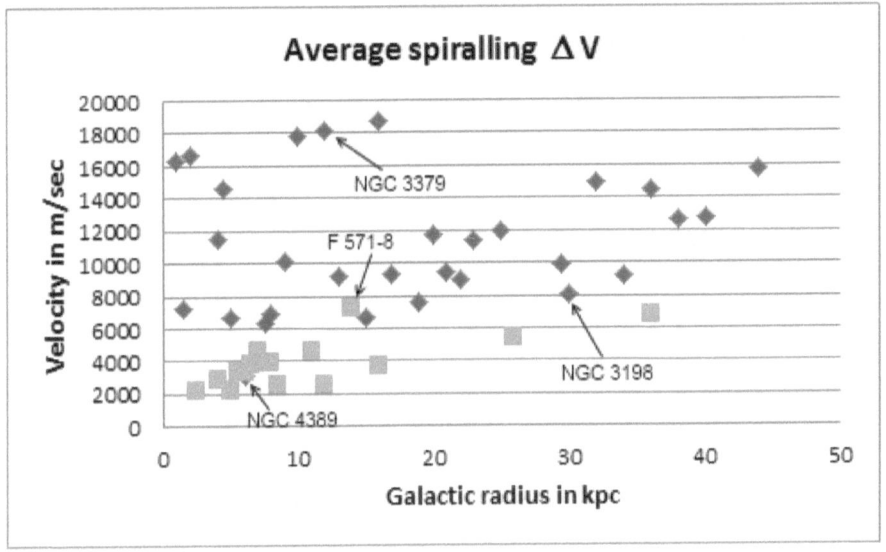

Figure 6.32 Average spiraling-in velocity
for low—and high-surface-brightness galaxies

The vertical scale of velocity varies from 2 to about 18 km/s on figure 6.32. It represents the average spiraling-in velocities for our forty-six selected galaxies, where $\Delta V = V_c - V_K$ is the average of all ΔV values for each galaxy. Again, diamond symbols represent HSB galaxies, and squares correspond to LSB galaxies. We observe a remarkable difference between HSB and LSB galaxies, where LSB values are generally much lower than HSB delta velocities and represent a much slower spiraling-in movement of orbiting bodies around LSB galaxies. LSB galaxy F 571-8 is located near the lower limit of HSB galaxies, but HSB galaxy NGC 4389, our special barred example, is oddly right in the middle of LSB galaxies due to a strong crossover that affects the ΔV averages. Elliptical galaxy NGC 3379 shows a large average ΔV of about 18 km/s, along with a few other galaxies where a relatively large nucleus could also be present.

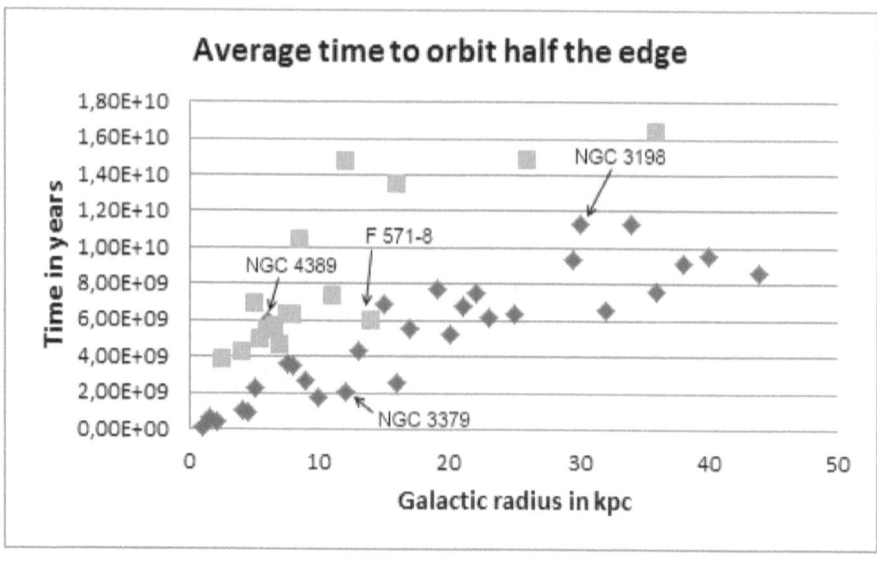

Figure 6.33 Average time to orbit a half-circumference distance

On figure 6.33, we find the average time for a body to orbit a distance equal to ½ the circumference at the edge of the galaxy. The vertical scale is time in years; diamond symbols are still related to HSB, and squares to LSB galaxies. Both LSB and HSB time values increase with the distance since time distance is directly related to the radius, but there is a clear cut in the distribution of LSB and HSB galaxies whereby the orbiting time taken by LSB galaxies is higher than for HSB galaxies. It varies from 4 to 16 billion years for LSB galaxies and implies slower orbiting, lower masses, and less activity as observed.

6.4   Orbiting Mass and Spiral Galaxies

Spiral and disklike galaxies are complex gravitational systems, with massive cores surrounded by a bulge that may extend to a bar before the spreading of the arms, followed by ill-defined nebulous material related to the surrounding wall. A sketch is drawn on figure 6.34 to help in understanding how a material may be spiraling-in from the outside to the inside of a galaxy and to show how, from mass-to-mass interconnections, centers of attraction may migrate to the center of mass.

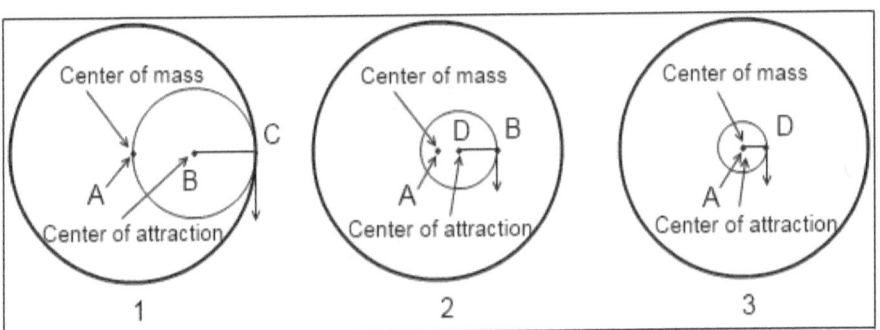

Figure 6.34 Migration of the center of attraction to the center of mass

Three circles representing the disk of a galaxy are drawn to shown how the center of attraction of different masses is gravitationally interconnected to the center of mass of the whole disk located at *A*, which is also the geometric center.

On disk number 1 at the left, we have an orbiting mass at *C*, whose center of attraction is located at *B*, halfway between *A* and *C*, so that the mass at *C* is moving with the corrected velocity, V_c, in the direction of the arrow and around the small circle because of the centripetal acceleration. Note that the center of attraction *B* is like a barycenter since for the mass at *C*, it is gravitationally acting as if the whole mass of the disk were concentrated at that point.

But on disk number 2, at the center of the figure, the mass that was located at the center of attraction at *B* in disk number 1 is now the orbiting mass around the center of attraction located at *D* so that mass *B* is orbiting with the corrected velocity, V_c, around the small circle. Note that as we go from orbiting masses at the edge to orbiting masses toward the center, the small circles decrease gradually in radius with the consequence that tangential movements draw a stronger and stronger curvature corresponding to a spiral.

Finally, on disk number 3 at the right, the mass that was located at the center of attraction at *D* in disk number 2 is now also the orbiting mass around the center of attraction located at *A* so that mass *D* is orbiting with the corrected velocity, V_c, around the small circle, and both the center of attraction and the center of mass are located at the same point *A*, which is also the geometric center. Since the displacements at *C*, *B*, and *D* are gravitationally connected and occur simultaneously, they participate in the overall rotation of the disk around the center of mass and correspond to the Keplerian velocity, V_K.

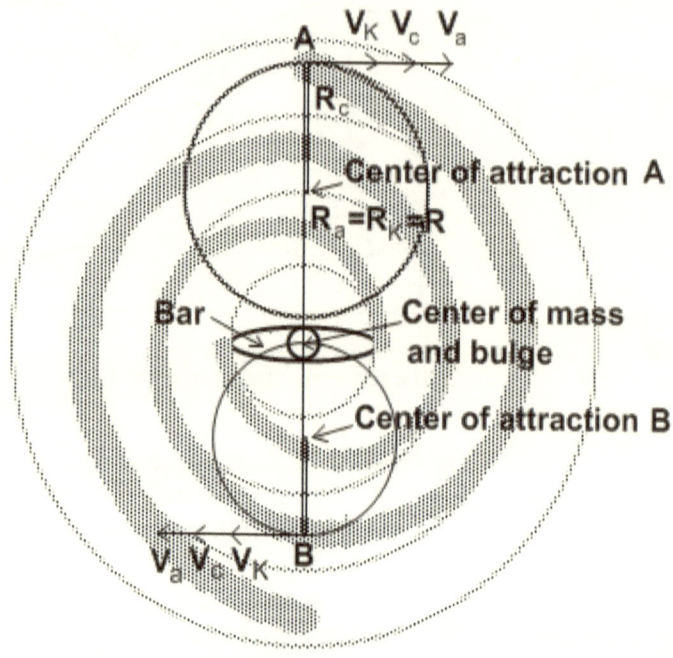

Figure 6.35 Sketch showing the relationship between an orbiting mass and a galaxy

A galaxy is drawn with central core, spheroid bulge, bar, and two gray-shaded spiral arms on figure 6.35. Four concentric circles are shown at equal intervals around the center of mass as a background reference. The bulge is the spheroidal shape near the center of the galaxy, where Kepler-Newton laws may approximately be applied. The bar is that part of a galaxy where rotation velocity between the bulge and the arms is equal to angular galactic velocity. Tangential velocity increases in the bar as the radius R. The bar is fed in fresh mass by the arms of the galaxy, and the oblate spheroid core, in turn, may be fed by the bar.

The mass present at A is associated with three velocity vectors whose lengths are proportional to velocity values. Keplerian velocity, V_K, is a spherical equivalent associated with center of mass radius $R_K = R$. Corrected velocity, V_c, is associated with radius R_c running from A to the center of attraction A. And the largest velocity, which is apparent measured velocity, V_a, is associated with center of mass radius $R_a = R$. So the center of mass radius is $R = R_K = R_a$.

A circle is shown around center of attraction A with radius R_c. This circle is much smaller than the center of mass circle (not drawn) passing through orbiting mass A. Centripetal acceleration from equation 6.26, where a = V_c^2/R_c = V_a^2/R, implies an equilibrium in mass acceleration toward the center of attraction and the center of mass. But the orbit of mass A is deflected by the stronger curvature associated with velocity V_c while rotating around the larger center of mass circle with velocity V_K that has a different acceleration.

The same observations apply to mass B, where a still-smaller circle is shown about 180 degrees down the spiral. The latter smaller circle implies a yet-stronger spiral curvature. Therefore, gravitating masses, around gradually smaller circles with smaller corrected radius R_c, will form a progressively stronger spiral curvature toward the galactic bar or central spheroid. But other less-significant compensating forces may be involved, like tidal effects creating a drag that could disturb the orbiting path. These effects will produce a spiral arm curving gradually more and more and funneling masses toward the center of the galaxy. So the spirals run from outside to inside and may be expressed as a decreasing function of radius R, some of them looking like a double Fibonacci spiral.

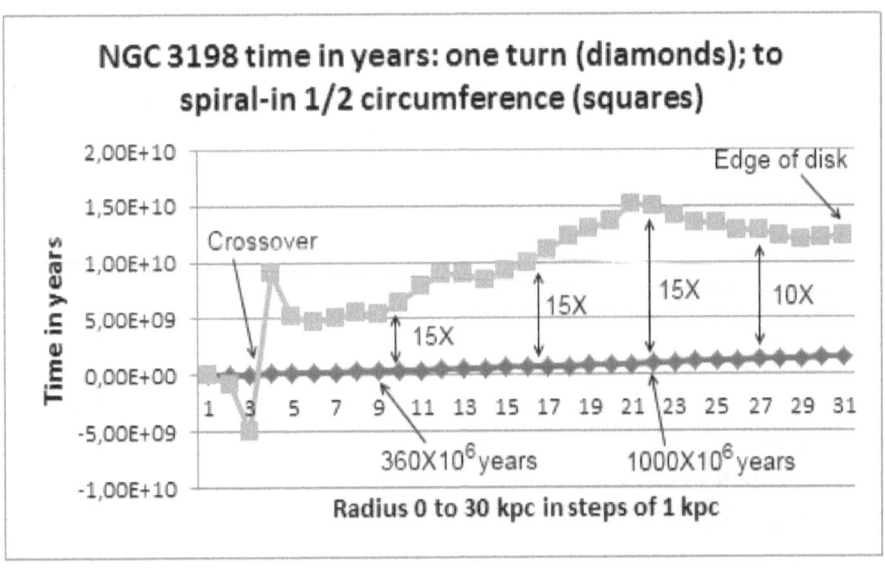

Figure 6.36 Distribution of orbiting time and spiraling time for NGC 3198

The distribution of orbiting time for one turn (diamonds) and spiraling-in time (squares) for NGC 3198 is shown on figure 6.36. The horizontal scale is the radius in kiloparsec, and the vertical scale is time in years. Orbiting time of 360 million years for one turn at 8 kpc is indicated as a comparison to our sun orbiting the Milky Way galaxy at about the same distance in a shorter time of 250 million years, and so NGC 3198 is a less massive galaxy with a larger radius (30 kpc versus 20 kpc) and without a bar. The crossover is forming an anomaly between the bulge and the arms. Orbiting time goes from 0 to over 1 billion years, and spiraling-in time for a distance equal to ½ the circumference of the ring varies mainly between 5 and 15 billion years. The most outstanding relationship is that it takes the same number of turns (about fifteen) to spiral in the same angle across most of the galaxy so that the spiral shape may be preserved for many billion years. But in the last third near the edge of the disk where spiraling-in takes less than fifteen galactic turns for the proportional distance, material should be falling in faster and stimulating star formation.

Orbiting masses with different radial values are not always moving at the same angular velocity in galactic arms since a galaxy is not a rigid disk. Angular velocity decreases with increasing R distance when a flat tangential velocity curve is present. Therefore, an orbiting mass near the edge of a galaxy will take a longer time to go down its spiral path (billions of years) since most masses are rotating with a small differential velocity proportional to $\Delta V = V_c - V_K$, which represents, for NGC 3198, about 11 billion years to go down a spiral equal to ½ the length of a circumference with a 30 kpc radius and for an average $\Delta V = 8.4$ km/s.

This indicates that most orbiting masses in every spiral galaxy are gradually and irremediably spiraling down toward the center. Thus, older stars must be found near the galactic center. So spiral galaxies seem to sweep and gobble down all material from their surroundings over billions of years, until exhaustion or merger.

A generalization of the basic model applied to nonsymmetrical fields of other celestial bodies should give a simple and more rational explanation of the advance of Mercury's perihelion since our sun is an oblate spheroid and has a measured flattening of about 6 km at each pole, which seems the right order of magnitude to produce the small advance in the planet's orbit. The same larger apparent velocity V_a versus corrected velocity V_c may apply, although to a lesser extent, to any planet-moon system or any multiple-star system since some flattening is generally present.

6.5   Linear Model for Filamentary Structures and Clusters

Galactic clusters and superclusters exhibit the same mass-to-light (M/L) ratio discrepancies, as discussed for single spiral galaxies, but with a much larger factor.

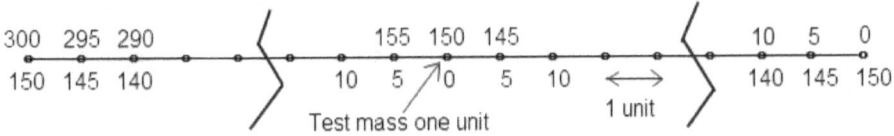

Figure 6.37 Linear model used for computations

The linear 1-D model of figure 6.37 shows a test mass of 1 unit located at the center of a 0 to 300 distance units line, which means at 150 units from both ends. So these are the same distances as the center and diameter of the 2-D disk model presented above. A sequence of 150 masses of 1 unit, all separated from the next one by a distance of 1 unit, is present on each side of the test mass. The test mass is moved in steps of 5 units to one end of the line, up to 150 units, for a total of thirty-one steps that are analyzed in a way similar to the larger disk of 150 units in the previous disk model but, in this case, without an extension of 50 units away from the end of the line.

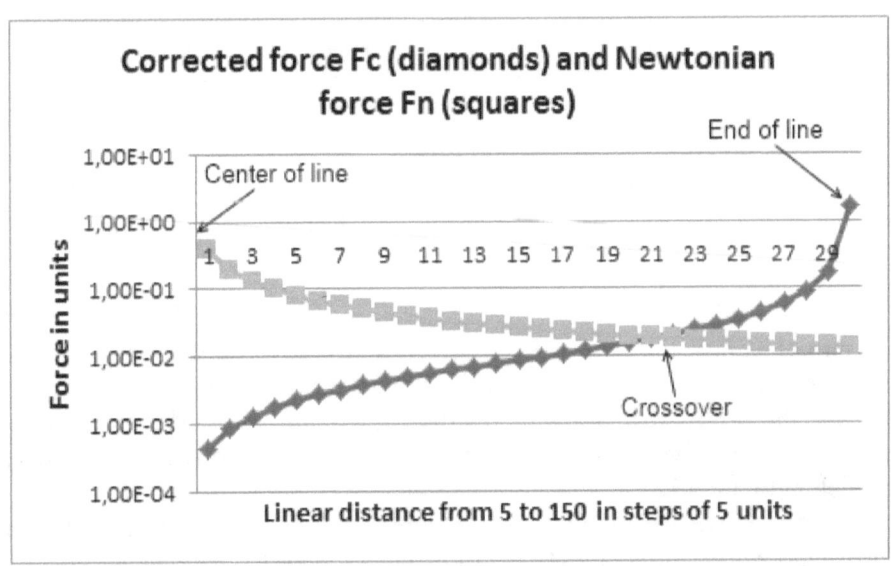

Figure 6.38, Corrected and Newtonian forces for a linear mass distribution

The horizontal scale runs from 5 to 150 units in steps of 5 units on figure 6.38 and no values are plotted for distance 0 since the vertical scale is logarithmic and cannot display 0. Gravitational force computations F_c are done for each separate mass and distance unit to obtain the progressively cumulative ΣF_c.

Force F_n is computed from the center of mass as the linear mass and distance increase on one side and decrease on the other side. From these, as for the previous disk model and with the same equations, mass and velocities will be calculated for an orbiting mass perpendicular to the line.

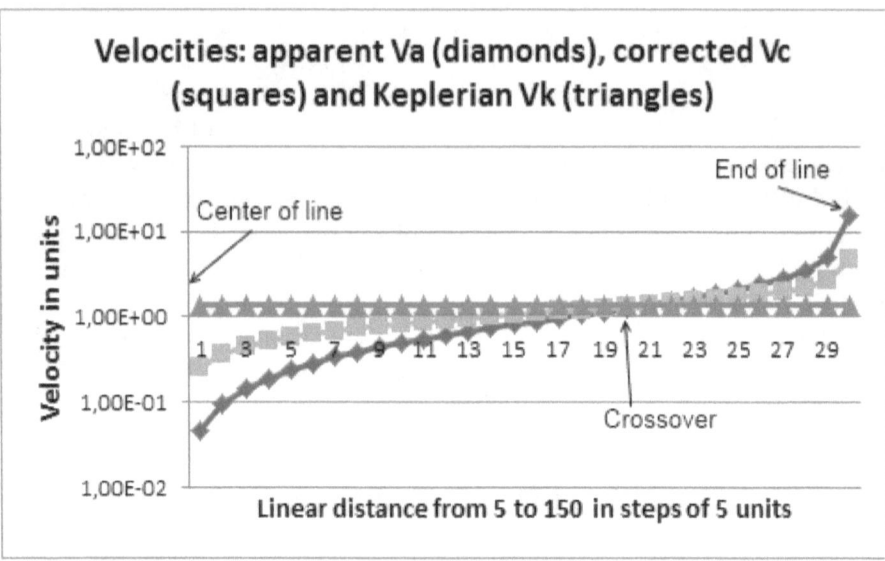

Figure 6.39 Apparent velocity, V_a (diamonds); corrected velocity, V_c (squares); and Keplerian velocity, V_K (triangles)

On figure 6.39, the horizontal scale also runs from 5 to 150 distance units in steps of 5 units. The test mass is moved from the center at the left and across the linear mass distribution toward the right. Vertical scale shows velocity in nonspecific units, with 0 absent on the logarithmic scale. We recognize a similar pattern of velocity increase with crossovers, as with the previous disk model on figure 6.12.

The curve with diamonds, ranging more than two orders of magnitude and crossing the other two curves to the right, corresponds to apparent velocity, V_a, which is an undervalued velocity on the left toward the center and overvalued toward the right end of the line. This is the kind of large velocity variation that could be measured from Doppler shifts in, for example, bars, linear galactic arms, linear strings of galaxies, or elongated groups and clusters of galaxies.

Squares represent corrected velocity V_c that will yield corrected mass when combined with corrected distance R_c in the circular velocity equation. The flat curve with triangle symbols corresponds to Keplerian velocity V_K that will also yield corrected mass when combined with center-of-mass distance R in the circular velocity equation.

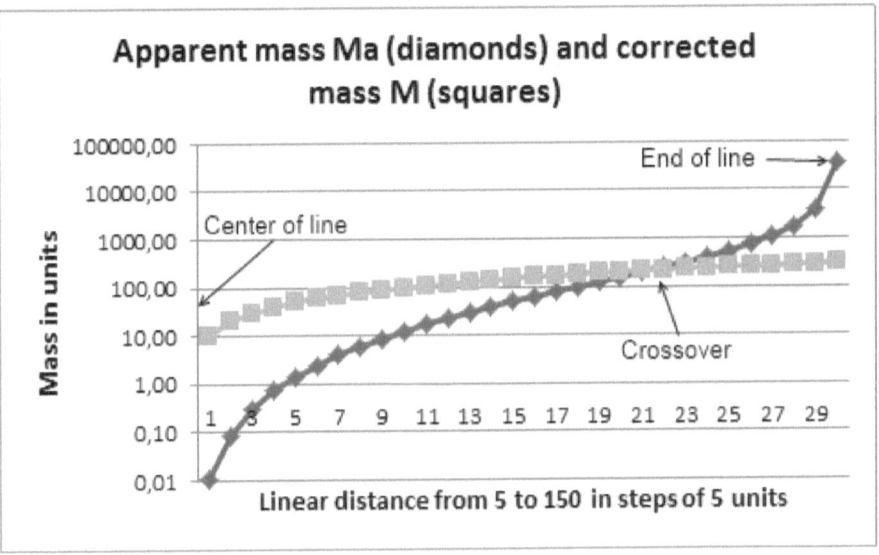

Figure 6.40. Distribution of apparent and corrected mass

The horizontal scale is the same as on the two previous figures, and the vertical scale with the 0 excluded is logarithmic and represents mass in nonspecific units on figure 6.40. The crossover at about 110 units separates, by about two orders of magnitude, the most significant and overvalued apparent masses to the right from the least significant and undervalued apparent masses to the left, and this all the way to zero. Corrected or real mass M (squares) varies from 0 to 300 mass units, but the first value shown on the log scale is 10 (2×5) since both sides of the line are added.

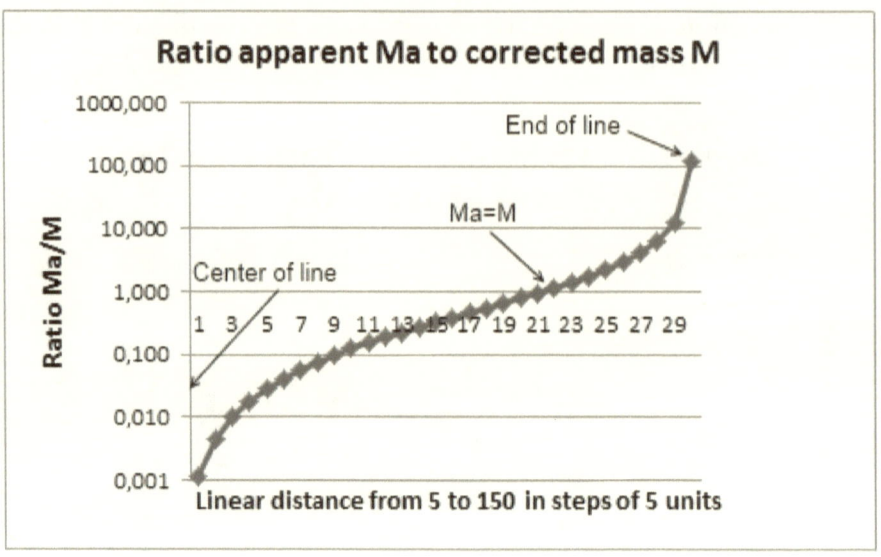

Figure 6.41 Ratio apparent to corrected or Keplerian mass M_a/M

The horizontal scale on figure 6.41 is the same as the previous figure and starts at a distance of 5 units. Vertical logarithmic scale indicates ratios between apparent and corrected mass M_a/M. The force and mass crossover is indicated, where $M_a = M$ at 110 distance units from the center of the line.

Thus, the curve represents ratio increases from the center to the end of the line between apparent mass M_a and corrected or Keplerian mass M, where the latter two correspond to real mass M. The increase is impressive since apparent mass is more than five orders of magnitude larger than corrected mass across a linear distance of 145 units from 5 units to the end. It means that mass could be underestimated across the least important three orders of magnitude near the center of such a linear pattern and overestimated across the two most important orders of magnitude toward the end of such a linear pattern.

For this linear model, we have an apparent mass 123 times larger than the corrected mass. This ratio is much more important than in the previous disk model, where the apparent mass was 4.7 times larger than the corrected mass for the theoretical model, and on average, M_a/M was 1.8 times larger for the forty-six selected galaxies. If the same difference is present for the linear model between theoretical and applied mass results, then we should find that apparent mass is forty-seven times larger than corrected mass on average.

Since bars, filamentary structures, some groups, and clusters are generally described as more or less elongated features, then this should be no surprise to find a large range of values for apparent-to-corrected-mass (M_a/M) ratios in field measurements, which would translate into a large range of mass-to-luminosity (M/L) ratios, much more than observed for disk-shaped individual spiral galaxies. This linear model may also be applied to some galaxies with elongated arms, such as NGC 6872 or UGC 10214, and in merging galaxies with trails of material. Finally, the linear model may be used to explain part of the excess velocities found in galaxies feeding clusters via the concave triangular ducts located between the walls and the clusters.

7. Four Forces or Facets of Energy in the Cosmos

Most books of physics and astrophysics state that four basic forces have been identified in the cosmos: gravitation, electromagnetism, the weak force, and the strong force. But the last three forces occur and are intertwined at the atomic level and may be grouped into a unique force—the electronuclear, which is described by the theoretical grand unification theory. So this would leave the two known basic forces, gravitation and the electronuclear, to which we may add a third one: the hypothetical black force that stops the collapse of massive black holes when all electronuclear forces have been exhausted. A fourth one may be suggested for the complexity bonds from subchemical to biochemical reactions. The complexity bond is this diffuse force driving reactions in the cosmos from the most initial and farthest subnuclear ones to the most complex biochemical and from which stem these cosmic qualities that we call cooperation and intelligence.

7.1 Electronuclear Force and Energy Conservation

The electronuclear force results from the interaction of all the fermions and bosons, as described in the table of fundamental particles in the Standard Model of physics. Since there are energy and particle exchanges at the quantum level and friction everywhere at the macroscale, thermodynamics dictates energy conservation and entropy in all processes.

7.1.1 Standard Model of Particles

The Standard Model of particles counts twelve fundamental elementary particles called fermions and five fundamental bosons, including the Higgs, as shown on figure 7.1. The figure represents an instant flash picture of the present-day knowledge of physicists that is continually moving forward to new confirmations and discoveries.

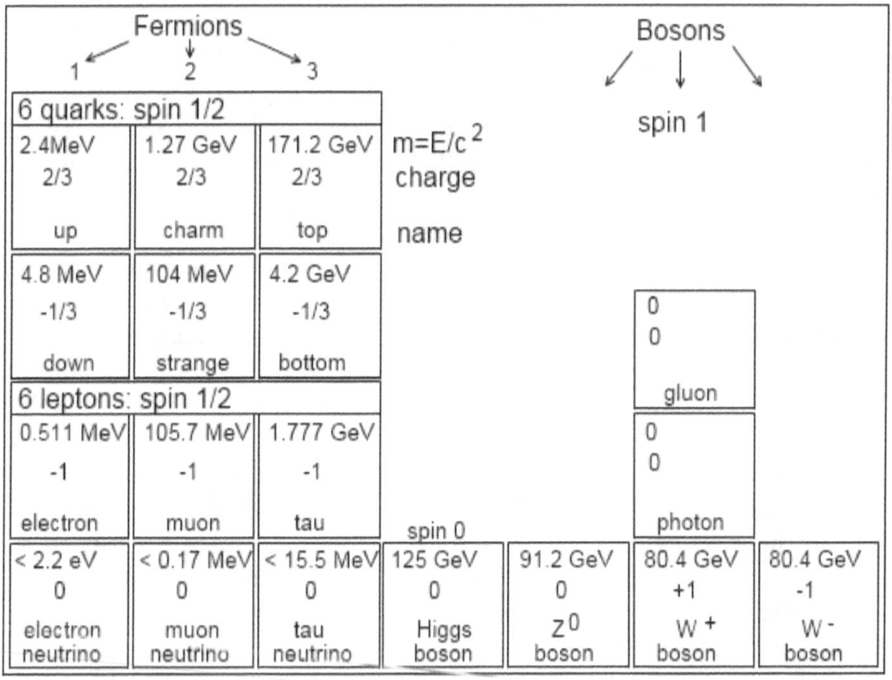

Fermions				Bosons			
1 ← 2 → 3				↙ ↓ ↘			
6 quarks: spin 1/2			$m=E/c^2$	spin 1			
2.4MeV	1.27 GeV	171.2 GeV					
2/3	2/3	2/3	charge				
up	charm	top	name				
4.8 MeV	104 MeV	4.2 GeV			0		
-1/3	-1/3	-1/3			0		
down	strange	bottom			gluon		
6 leptons: spin 1/2					0		
0.511 MeV	105.7 MeV	1.777 GeV			0		
-1	-1	-1			photon		
electron	muon	tau	spin 0				
< 2.2 eV	< 0.17 MeV	< 15.5 MeV	125 GeV	91.2 GeV	80.4 GeV	80.4 GeV	
0	0	0	0	0	+1	-1	
electron neutrino	muon neutrino	tau neutrino	Higgs boson	z^0 boson	W + boson	W - boson	

Figure 7.1 Fermions and bosons of the Standard Model

The twelve fermions are subdivided into six quarks and six leptons, with spins ½. Each of them is represented by a little square, where the first value on the top is the energy of the particle, which may be divided by c^2 to obtain the mass. It is followed below by the charge of the particle, which corresponds to uneven values of $2/_3$ and $-1/_3$ for the quarks and to even values of -1 or 0 for the leptons. The name of the particle is given at the bottom of each square. There is a crescendo in the mass of the particles from the first generation in the first column on the left to the third column on the right. Most of the known cosmos is composed of the up and down quarks forming nuclei, which are combined with electrons to form atoms. The photon is the exchange boson that we are most familiar with, but the unfelt bosons are just as important in the nuclear reactions, together with the flow of neutrinos.

The five bosons are stranger since two of them, the photon and gluon, are considered massless and without a charge, and the other three, Z^0, W^+, and

W, have masses of 91.2, 80.4, and 80.4 GeV and charges of 0, +1, and −1, respectively. The latter three are the exchange bosons for the weak force or interaction in beta and other nuclear decays and reactions.

Photons represent transient energy absorbed or emitted by particles both at the electronic and nuclear levels. Their energy and momentum depend only on frequency, and they are related by the equations $E = hf = pc$, where h is Planck's constant, f is frequency, p is momentum, and c is their velocity in a vacuum. Photons have no rest mass in the Standard Model; they carry the electromagnetic force and behave both like waves and elementary particles described as discrete quanta of energy.

Gluons are, like photons, exchange particles carrying energy between different types of fermions at the nuclear level. They carry the strong nuclear force between quarks and have no rest mass in the Standard Model. No isolated gluon has ever been studied, and no quark either, since both are glued so strongly together with a force that increases with distances in the nucleus; no particle accelerator has succeeded yet to separate a single gluon by breaking this link.

The Higgs boson is an elementary particle under study by some of the teams at CERN near Geneva. It is an exchange particle that is proposed to explain the mass of most particles and the absence of mass for two of the bosons. In 2012, CERN scientists announced the discovery of a particle compatible with the Higgs boson, but much work is still needed to confirm the attributes of the particle.

Except for the photon considered as its own antiparticle, for every particle, there should be an antiparticle born in high-energy pair production, but that kind of symmetry obviously does not exist since the cosmos, as we experience it, is almost made of particles only. So somewhere in the process of particle production within the maxibang events, the symmetry is broken and particles emerge as the dominant half of the pair. Matter has survived the antimatter annihilation because some unknown subparticle or sub-boson has intervened and broken the symmetry early in the process of bubble expansion. The experimental proof has not yet been demonstrated.

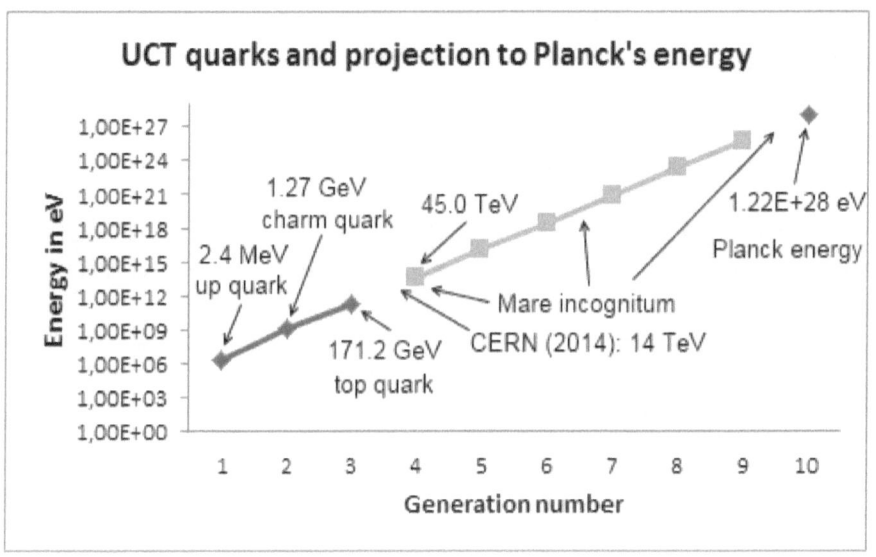

Figure 7.2 UCT quarks and projection to Planck's energy

Up, charm, top (UCT) quarks and projection up to Planck's energy are shown on figure 7.2. The horizontal scale represents generation or family number, and the vertical scale is energy in eV with logarithmic divisions in multiples of three orders of magnitude.

The first three values of 0.0024, 1.27, and 171.2 GeV correspond to the three quarks with charge $2/_3$ in figure 7.1. The last value of 1.22×10^{19} GeV represents Planck's energy from the big bang model. It would be located at the gravity wall, where the electronuclear energy would not cross but would rebound if transposed to a maxibang model. All the squares in between are approximately distributed with a constant interval, and they represent the *mare incognitum* of particle physics. The first square corresponds to a level of energy of 45 TeV while the Large Hadron Collider at CERN is expected to reach only 14 TeV in the year 2014.

Now, imagine all the new particles and families of particles that could lurk in *mare incognitum*. Thus, it seems research and new discoveries await the future generations of humanoids. The task of scientific discoveries is far from being completed since we have no information covering most of the expected energy range in the cosmos. Beyond the known electronuclear, we should find a high-energy realm that will explain the early maxibang expansion at superluminal velocities and also a most likely new subquark force, the black force, that halts the collapse of black holes to within finite values, and this all the way to the instant before the maxibang implodes and explodes from the most massive black hole containing the equivalent mass of about seven thousand galaxies at the critical point or the equivalent of 171 trillion sun masses.

7.1.2   Friction, Entropy, and Energy Conservation

Tectonic plates creep and shake, wheels squeak, and atoms heat up when squeezed, so friction is present everywhere, and the pure mathematical laws of physics are plagued with imperfections because we live in a dynamic electronuclear domain of the cosmos where movement is sparked by the exchange of energy. So the four laws of thermodynamics apply equally to all electronuclear processes in expanding bubbles, in galaxies, in clusters, and in quasars.

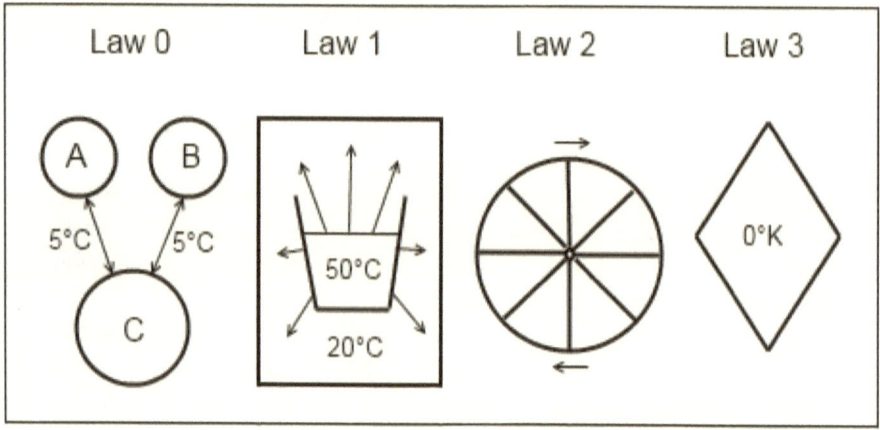

Figure 7.3 Sketch of the four laws of thermodynamics

Law number 0 of thermodynamics simply states that when two systems, *A* and *B*, are separately in thermal equilibrium with a third one, *C*, then they both are in thermal equilibrium with each other. Translated in terms of temperature, this means that we can measure and compare temperatures from different systems. This is currently used in astrophysics when radiations from stars or galaxies are compared to establish their level of activity.

The first law is the best known and easy to understand: it states that energy is conserved in all processes since nothing is lost, nothing is created, but everything is transformed. Thus, heat and work have only the effect of adding or removing energy from a closed system without affecting the total energy balance sheet. This law implies that a perpetual-motion machine is impossible since an input of energy is necessary to do some work. On figure 7.3, a vessel containing a liquid at 50°C radiates its energy to the surrounding room at 20°C, so heat is only transferred to the room until thermal equilibrium is reached.

The second law is well-known and often cited but more difficult to grab since it nearly always represents the increase of a negative factor in energy exchanges. The law states that the entropy of an isolated system never decreases but that it almost always increases so that such a system evolves toward a state of maximum entropy while reaching thermal equilibrium. Since converting energy to work without frictions or any other side effects is impossible, then building a perpetual-motion machine that would convert endlessly one source of energy to work is also impossible. The wheel on the figure is flawed with friction, so its kinetic energy is bound to decrease gradually while increasing entropy.

The third law deals with the entropy of a system at low temperature. It states that the residual entropy tends to a constant value as the temperature of the system tends toward zero, and thus, at the limit in the classical interpretation, the entropy of a pure crystal is zero at the absolute zero and represented as 0 on the Kelvin scale. The law is symbolized with a diamond at 0 K on figure 7.3.

7.2   Gravitation and Electronuclear Rewinding

As already mentioned, the force of gravity plays a unique role in the cosmic rewinding of all the electronuclear energy. Since gravity is a conservative force, it has special attributes not found in the electronic or nuclear forces. The speed of the gravitational interaction has never been measured, but it must be more than constant c since gravity is not confined within the black holes.

7.2.1   The Conservative Force

In classical physics, when the work done to move a particle from point A to point B is independent of the path followed, then the force is conservative. If the particle moves in a loop from A to B and back to the same kinetic energy in A, then the total work is zero for a conservative force. Gravity is a conservative force that depends only on the position of the object to which a potential value may be assigned. On the other hand, friction represents energy losses in the form of heat and is an example of a nonconservative force.

Figure 7.4 The fate of Sisyphus rolling forever a boulder up the hill

In Greek mythology, Hades was lonely as a god-king in the underworld across the river Styx. He abducted Persephone, the daughter of Zeus, and made her his wife, so they jointly welcomed their earthly human clients. Sisyphus, who had reigned as king of Corinth, was a new client for Hades but also a cunning artist who handcuffed the latter so that he could keep on partying on earth. This action created havoc in the earthly entropy since nobody could cross the Styx anymore. Hades was finally liberated, but Sisyphus tricked Persephone and borrowed an extra time of pleasure on earth. Such a crime against a god and a goddess was severely punished: he was condemned for eternity to roll a huge stone up to the top of a hill with a tremendous effort just to see the stone roll back to the bottom and then start all over again.

This story illustrates how the force of gravity does not rub off or wear out no matter how often Sisyphus repeats his absurd task. The potential gravitational energy is always present with the same intensity at the top of the hill, and the stone acquires kinetic energy as it rolls downhill while the potential energy decreases.

7.2.2   Gravity and Velocity

What is the velocity of gravitational interconnections between two units of energy in the cosmos? And what is the velocity of the effects of the gravitational force on the electronuclear particles? These are two of the most fundamental questions in our quest to understand the cosmos [50].

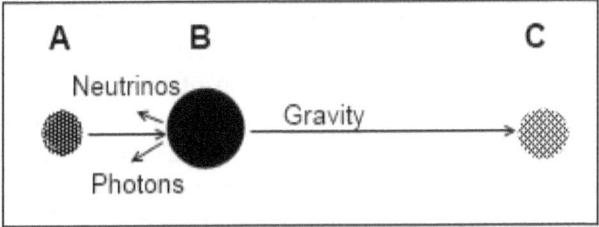

Figure 7.5 Gravitation is the deepest and most overpowering of all forces.

Although gravitation is the most familiar and inescapable force of nature, it is also the most remote and out of reach with present-day technology. Figure 7.5 shows three bodies, *A*, *B*, and *C*, lined up and interacting. Electronuclear energy emerging from *A* will be absorbed or reflected by the massive neutron star located at *B* and so will also be neutrons and neutrinos. But the gravitational field influence of *A* is fully felt by every neutron in *B* and also by every particle in *C*. The presence of *B* does not alter or deviate in any way the gravitational influence of *A* on *C*. In other words, we could remove *B* and the gravitational force between *A* and *C* would be exactly the same. No experiment has ever proven the contrary.

Moreover, there is a physically measurable delay for photons and neutrinos reaching *B*. The delay is proportional to the speed of light, *c*, for photons and slightly less for neutrinos, but no serious experiment supporting the propagation of gravity at light's speed has been demonstrated in the past century, and no measurable aberration related to propagation delay may be observed for gravity between *A* and *B* or *C*. Gravity is the permanent physical contact between every two units of particle-energy in the cosmos that obeys to the inverse square law in one dimension. But the effects of gravity on the electronuclear are measured and perceived at a maximum speed of *c* since photons emitted in such an interaction cannot travel faster than light.

The speed of gravity is not the speed of light or an infinite speed, but it is simply a permanent instantaneous-like contact in a cosmos with finite dimensions. In other words, two units of energy are always within a finite reach of each other. Otherwise, it would be utter chaos in the cosmos without the cohesive force of gravity that glues together all particles within the 286 million l-y diameter of the spheroid. For example, two electrons with a mass of 9.11×10^{-31} kg each and separated by the cosmic diametrical distance of 2.71×10^{24} m are gravitationally attracted to each other by an extremely weak force of 7.55×10^{-120} Newton. Over cosmic distances, forces and densities are so low that there is no risk of collapse.

The gravitational field notion covers the effects of a large number of energy units, like within a proton or a star. The theory of general relativity from Albert Einstein is a special description of the time-space behavior of electronuclear particles in a cosmos under the control of gravity. It is special relativity that brought the confusion about the speed of gravity by stating that nothing can exceed the speed of light. But general relativity uses instantaneous coordinates and momenta in its equations of motion, thus implicitly recognizing faster-than-light gravity. Theoretical physics has continued to strive toward a theory of everything, which would unify the Standard Model with quantum gravity and large-scale relativity [71] [79]. The most promising attempt seems to be loop quantum gravity, but as we have seen, the scale of energies is so tremendous that no machine will be able to probe to the core of gravity in the foreseeable future unless we find a way to extract useful information from the black holes.

7.2.3 Cosmic Cyclic Rewinding

We will review the time sequence computed and projected by big bang proponents and interpret this sequence as a maxibang event with its peculiarities. Then, we will use Planck's length as the radius of a sphere, compute the volume, and find the density and other parameters for a compacted or expanding mass energy equivalent of about seven thousand galaxies. Finally, we will briefly describe a heat pump and apply its physical characteristics to the cosmic self-rewinding heat pump.

Bang time sequence for our bubble				
Time	Time in seconds	Energy in eV	Temperature in K	Event
13.8 billion years	4.35E+17	2.59E−04	2.73E+00	Present age of our bubble
1.0 billion years	3.16E+16	6.29E−03	1.00E+02	Galaxies form around our bubble
380,000 years	1.20E+13	2.59E−01	3.00E+03	Radiation starts to escape
70,000 years	2.21E+12	8.62E−01	1.00E+04	Matter dominates radiation
1.0 year	3.16E+07	8.62E+01	1.00E+06	Photons dominates growing bubble
1.0 hour	3.60E+03	8.62E+03	1.00E+08	Formation of prime helium
1.0 second	1.00E+00	8.62E+05	1.00E+10	Neutrinos start to escape

1.0 millisecond	1.00E–03	8.62E+07	1.00E+12	Formation of hydrogen nuclei
1.0 microsecond	1.00E–06	4.31E+09	5.00E+13	End of quark domination
1.0 nanosecond	1.00E–09	8.62E+10	1.00E+15	Quark dominates in expanding bubble
1.0 picosecond	1.00E–12	4.31E+12	5.00E+16	Weak and electromag forces separated
1.0 femtosecond	1.00E–15	8.62E+13	1.00E+18	Quark and antiquark formation
End strong phase	1.00E–35	8.62E+23	1.00E+28	Separation of strong and electroweak
Planck time	1.00E–43	1.22E+28	1.42E+32	Electronuclear rebound on gravity wall

Table 7.1 Bang time sequence for our bubble

The bang time sequence and follow-up events in table 7.1 are transposed from the big bang to the maxibang at the origin of our bubble. The first column indicates the main sequential time scale of events, which is expressed in seconds in column 2. Then, related energy and temperature are respectively shown in column 3 and 4, with temperatures varying from 1.4×10^{32} to 2.73 K. Events are shortly described in column 5 from Planck time to present day. One problem with the big bang model anchored in Planck units is that the Planck mass is only 2.18×10^{-8} kg, about the mass of twenty-two thousand amoeba, quite short of the cosmic mass [72].

The first and most remarkable difference from big bang interpretation is that in a maxibang, at Planck time, the electronuclear material emerges from the quasar implosion as a rebound on the gravity wall, and the rebound is not limited by the velocity of EM waves since the emergence is from a different physical domain. So before the electronuclear material escapes, the black force and subnuclear particles dominate the expansion of the shell at a velocity greater than c until a bubble size is reached, where the mass of the shell can escape from collapse at a lower velocity than c. This phenomenon has been described as cosmic inflation in the big bang paradigm whereas it is a gradual expansion at superluminal speed in the maxibang model. The second remarkable difference is that 13.8 billion years is the age of our own bubble and not the age of the cosmos [83].

As surely as massive stars collapse to neutron stars and black holes, massive clusters of one thousand galaxies plus the equivalent gas mass of six thousand galaxies, likely dressed as quasars with a giant black hole in the center, should collapse all the way to the final force of gravity since all other forces, including the electronuclear and subnuclear black force, have been dissolved and merged under the effect of gravity. Then, in a gravity rebound, it is possible to expand faster than the speed of light to a bubble volume and mass where the electronuclear takes over for a standard expansion at a shell radius equal to or greater than 53.2 l-y or 16.3 pc. But we will see below that the standard expansion may start with a shell radius one hundred times larger if the whole mass of the cluster is included in the equations.

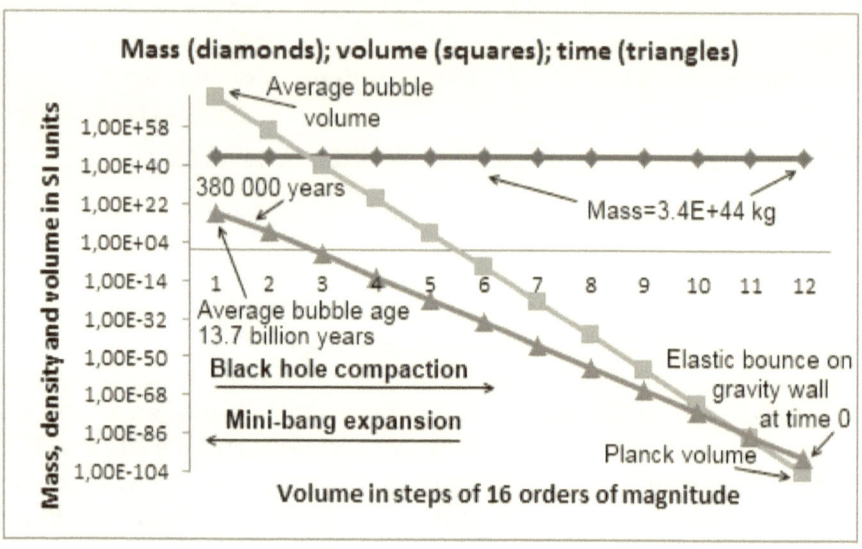

Figure 7.6 Black hole and maxibang mass, volume, and time distribution

The following three graphs are based on one Planck unit only: the Planck length, which is 1.6162×10^{-35} m and is applied as the radius of the smallest and most compact sphere that may be found in nature, so the minimum volume is equal to 1.77×10^{-104} m^3. On all three graphs, we show that black hole compaction proceeds to the right with smaller volumes and that maxibang expansion runs to the left with larger volumes although discontinuities are not inserted in the curves.

The graph of figure 7.6 shows volume and time distribution in the presence of a constant mass of 3.4×10^{44} kg corresponding approximately to the critical mass in our maxibang expansion and contraction model. The horizontal scale for volume is divided in twelve steps of sixteen orders of magnitude each, and the logarithmic vertical scale presents SI units for mass, density, and volume.

The square symbols represent bubble volume variations in m³ over one half cycle. If the first half cycle is bubble expansion to a maximum of 1.77×10^{72} m³, then the other half is the contraction of material borrowed from common bubble walls and gathered to form a giant cluster of about one thousand galaxies plus massive amount of gas that will be squeezed to the Planck volume. The curve with triangles corresponds to a time of 4.32×10^{17} seconds at the extreme left, which is equal to a bubble age of about 13.7 billion years, the expected age of our bubble. A milestone marking the birth of the electronuclear domain that is shortly followed by the escape of photons is shown at 380,000 years. Then, time values decrease to zero at the extreme right, where compression reaches the uttermost value at the instant when the bang paroxysm occurs.

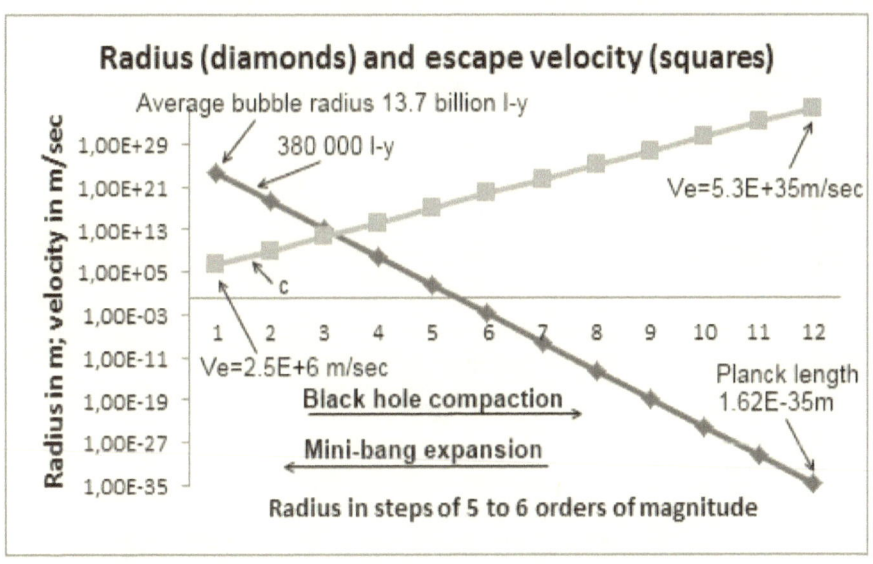

Figure 7.7 Black hole and maxibang radius and escape velocity distribution

On figure 7.7, decreasing radius and increasing escape velocity curves are presented with a horizontal scale divided in twelve steps of five to six orders of magnitude each for radius values shown with diamond symbols. The first value on the left represents an average bubble radius of 7.5×10^{23} m or 13.7 billion l-y, followed by the critical bubble shell radius of 380,000 l-y, marking the end of the dark force's and subquark particles' domination. The last value to the right is the Planck radius, with a length of 1.62×10^{-35} m. In between the extreme, we find all the intermediate radial values observed and hypothesized during the expansion-contraction cycle.

The escape velocity curve is shown with square symbols and start at the extreme left with a value of 2,500 m/s, which is the average radial expansion velocity for bubbles. It is shortly followed by constant c, the speed of light in a vacuum, and it increases to the extreme right to a maximum value of 5.3×10^{35} m/s at time 0. Of course, such high velocities are in the superluminal range and must be explained. In the contraction phase of the cycle, as soon as massive stars generate black holes, we find part of the material beyond escape velocity c, and as the black holes grow, the subnuclear particles and black forces intervene to form plateaus in black hole contractions [58]. In the expansion phase, the shell of the expanding bubble must reach a radius of 5.07×10^{19} m or about 5,350 l-y before the electronuclear material flashes out of the black force and subnuclear particles and sheds light on its neighboring part of the cosmos.

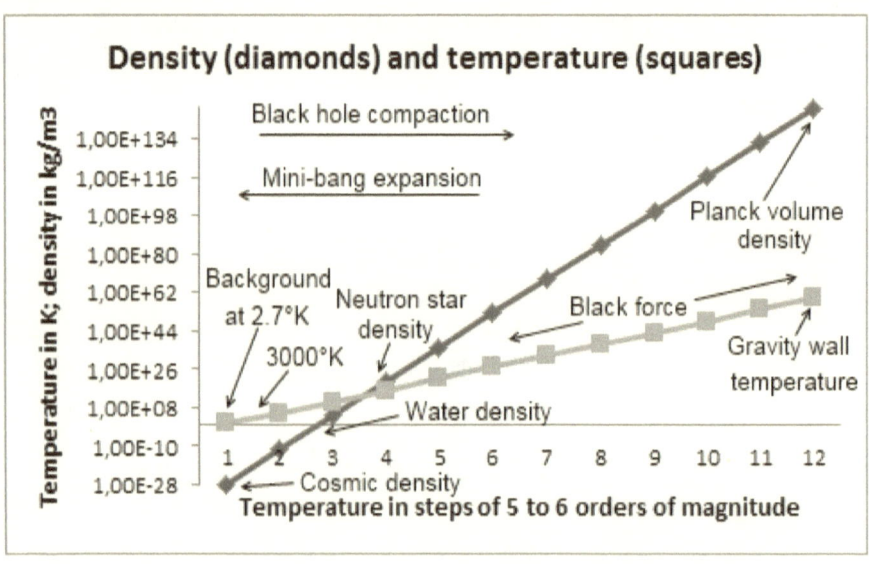

Figure 7.8 Black hole and maxibang density and temperature distribution

Density and temperature distributions are sketched on figure 7.8, with a horizontal scale divided in twelve steps of five to six orders of magnitude each for the temperature. Black hole compaction runs from left to right and maxibang expansion from right to left. Densities are shown with diamond symbols, which are starting at the extreme left with the average cosmic density of 0.21 M_p/m^3, followed by water and neutron star densities indicated as milestones. The last value to the extreme right is 1.92×10^{148} kg/m^3, the density calculated for the critical maxibang mass and the Planck volume.

The curve with square symbols represents temperature variations, with the first value at the left corresponding to the cosmic background temperature at about 2.73 K, which is followed by the average temperature of 3,000 K out of the black age. The last value to the right corresponds to gravity wall temperature at 2.38×10^{59} K at which all the compacted material from the equivalent of 171 trillion solar masses will rebound to form a new bubble. The black force should be present across most of the curve since densities exceed the escape velocity of the electronuclear material. Note that our original tiny and extremely compact sphere rapidly becomes an expanding shell where high densities and temperatures are maintained in the thin volume of the shell, so we expect some deviations from the projections for both contraction and expansion models.

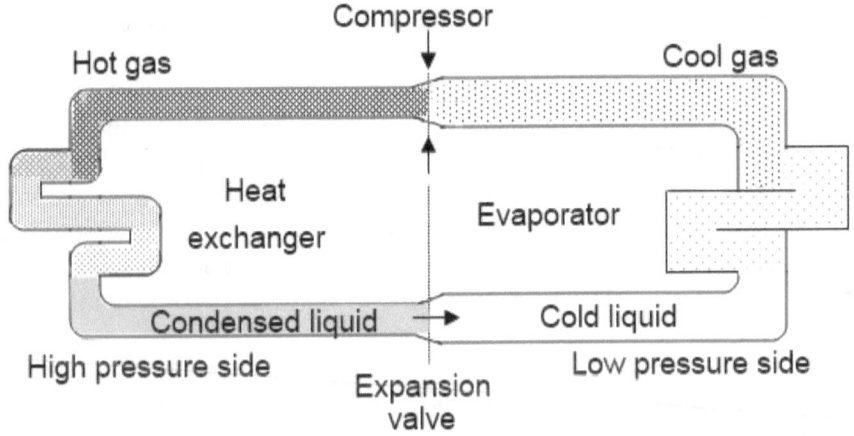

Figure 7.9 Sketch of a heat pump with cold and hot parts

A sketch of a heat pump is presented on figure 7.9 with its cold and hot sections. The idea is to compare a heat pump cycle to compression and expansion found in a cosmic cycle. In a heat pump, we find a compressor heating up cool gas to a higher hot gas temperature. A heat exchanger at high pressure extracts some of the heat from the gas for heating purposes. The gas is transformed to a colder condensed liquid that is fed into an expansion valve from which it emerges as a cold and lower pressure liquid. The latter is evaporated to form a cool gas that will restart the cycle through the compressor.

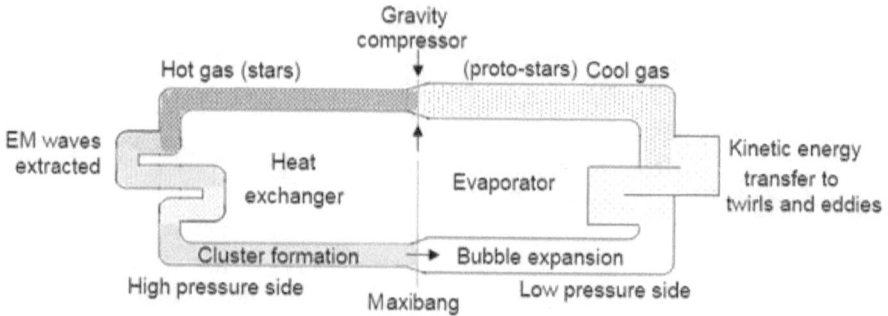

Figure 7.10 Sketch of the self-sustained cosmic heat pump

The sketch of the self-sustained cosmic heat pump is presented on figure 7.10, with a transposition of the same elements. The cool gas is found in the small and giant molecular gas clouds generated in bubble expansion and mostly associated with spiral and irregular galaxies. The cool-to-hot gas compressor is the force of gravity that will produce stars (hot gas) from the collapse of protostar eddies and twirls in the cool gas. Hot stars produce in their high-pressure interior a large range of EM waves, including heat released to the surrounding area, in a process similar to that of a heat exchanger. Additional heat is released with cluster formation and black hole compaction while maximum entropy is reached.

The maxibang at time 0 with zero entropy is the turning point between the high—and low-pressure sides of the cosmic pump since it marks the start of bubble expansion and cooling accompanied by a gradual increase of the rate of entropy across the whole cycle. The kinetic energy from bubble expansion is transferred to twirls and eddies when the expanding gas meets older walls or other bubbles to form galaxies equivalent to cool gas in a process similar to that of an evaporator.

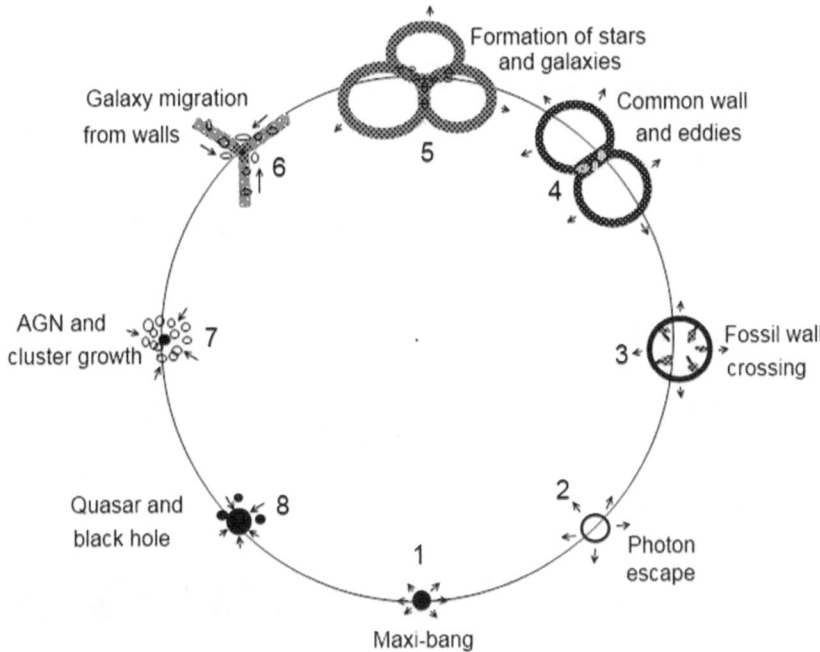

Figure 7.11 One complete cycle from maxibang to black hole

A sketch in eight steps of a complete cycle from maxibang to black hole with steadily increasing entropy is presented on figure 7.11. The first step is the maxibang event at time 0 that is followed by expansion at superluminal speed. A few thousand years later and at sublight speed, photons escape from the expanding shell. In the third step, fossil walls are crossed by the growing bubble that will also, in the fourth step, reach other expanding shells to form common walls and eddies at the origin of stars and galaxies. The bubble forms, in the fifth step, up to twelve common walls and up to 200 groups of galaxies with neighbors. Then, in step 6, galaxies and gas migrate from the walls to clusters while merging. In step 7, clusters are growing and merging around the increasing mass of a giant central galaxy with an active nucleus that will be gradually transformed to a quasar harboring a supergiant black hole in step 8. The supergiant black hole will implode under the force of gravity and a new cycle will start.

In a complete cycle, black holes are born early from massive stars, and they gradually grow to huge masses within and at the center of galaxies from which they merge into still-larger active-core galaxies and quasars until the final union of about one thousand galaxies plus associated gas that will collapse at a speed of about 5.3×10^{35} m/s in a single black hole that produces the bang.

7.3 Complexity Level and Energy Quality Conservation

In this final portion, we will have a brief look at the development of more and more complex structures that we observe in the cosmos, and we will question the added value of all the new qualities generated at every new level of complexity. In other words, are these qualities part of a new type of force and energy related to some cosmic network based on fundamental laws of nature, or is it some kind of dissipative energy?

7.3.1 Complexity Levels from Gravity to Cosmic Community

So the question is, are we generating conservative or dissipative energy when quarks unite to form protons or neutrons or when molecules unite to form proteins and when organisms unite to form society? It boils down to asking, is the whole greater than the parts when quality and efficiency are increased at a higher complexity level?

Complexity level	Force or bond	Comments
Gravity wall	Gravity	Final encounter at the bang
Subquark	Black	Superluminal and out of experimental reach
Quarks	Strong	Exchange bosons are the gluons
Neutrons/protons	Weak	W and Z bosons and neutrino particles
Nucleus/electrons	Electromagnetic	Electronic charge and spin
Atoms	Chemical bonds	Covalent, ionic, and metallic bonds
Molecules	Molecular bonds	Hydrogen and dipole interactions
Proteins	Peptide bonds	Covalent, hydrogen, and mix bonding
DNA/RNA	Phosphodiester bond	Covalent bond with enzyme-catalyzed reaction
Cells	Biochemical bond	A mixture of weak and strong bonds
Organism	Cellular bond	Cells forming mechanical/biochemical bonds
Society	Cooperative bond	Group survival by product/info exchange
Cosmic community	Sharing bond	Superluminal exchange and travel

Table 7.2 Distribution of growing complexity in cosmos

The distribution of growing complexity in the cosmos is presented in table 7.2, where the first column is the complexity level, the second column shows the corresponding force or bond, and the third column gives a brief comment.

The first level is gravity at the zero and near-zero time when all the future materials of one bubble are accelerated down the final black hole at a superluminal speed and rebound on the gravity wall where the entropy is zero. The second level is also in the superluminal range of velocities and out of reach for our limited present-day particle accelerators. It involves the black force acting on subquark particles. The third level represents the quarks united by the strong force with the gluons acting as boson exchange particles.

At the fourth level, we find the W and Z bosons as the exchange particles of the weak force for the emission and absorption of neutrinos in and out of neutrons. The next level is the electromagnetic force associated with the exchange of photons by electrons and other particles.

The following levels are all electromagnetic chemical-to-biochemical bonds. Between atoms, we find the covalent, ionic, and metallic chemical bonds, which vary in strength. Molecular bonds linking two or more molecules are generally related to hydrogen and dipole interactions. Proteins are formed by peptide bonds where covalent, hydrogen, and mixed bonds are present. The DNA and RNA photocopy molecules are formed by phosphodiester bonds, which are covalent bonds catalyzed by an enzyme (protein).

Living cells are born from a mixture of weak and strong biochemical bonds, including proteins and DNA/RNA. Cells are united to form coordinated organs and the moving mechanical parts of an organism. Many individual organisms may join together to form a society where forces or bonds are directed to a cooperative bonding that yields product and information exchange for the survival and well-being of the group. Finally, many planetary or other cosmic societies may unite in order to share information and use common luminal and superluminal exchanges, including transport, for reciprocal betterment.

We know that the entropy of the electronuclear material increases constantly with time even when we go up the rungs of the complexity ladder. Simultaneously with structure increase, we find a general increase in the exchange of information between different units of a specific complexity. It is this data that becomes a plus, contributing to the enrichment of the cosmic network of information. Thus, for a while, some energy is transformed and maintained as bits of qualitative information although at the end, any unit or network is faced with the inescapable increase of entropy.

7.3.2   Quality Increase and Conservation

We will look at quality increase for a specific example when we go from a simple to a more complex level in cosmic bubble evolution. It will be followed by a discussion of what is conserved from all the extra qualities issued from the increased complexity of the cosmos.

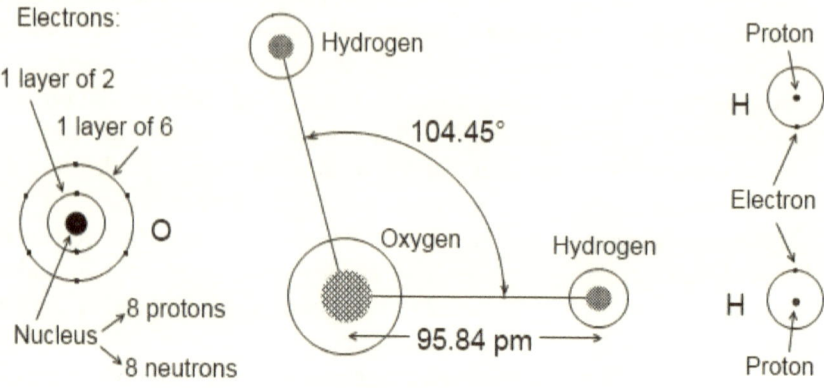

Figure 7.12 Water molecule formed from three atoms

A water molecule formed from three atoms is presented in the center on figure 7.12, with a separate sketch of the two hydrogen atoms on the right and one oxygen atom on the left. The hydrogen atom is the simplest of all atoms and may be represented as a nucleus containing a single proton that is surrounded by an orbiting electron with a charge opposite and equal to that of the proton. When two hydrogen atoms unite, they form molecular hydrogen in which the two electrons are equally shared in a covalent bond in equilibrium with the charge of the two protons. The oxygen atom has eight protons and eight neutrons held together by the strong force in its nucleus. The charge of the eight protons is neutralized by the opposite charge of two layers of electrons: one inner saturated layer of two electrons and one unsaturated outer layer of six orbiting electrons.

Since saturation of the outer layer would require a total of eight electrons, then there is room for sharing this layer with two electrons from other atoms. This is what is represented by the central sketch of a water molecule, where two hydrogen atoms are each sharing their unique electron with an atom of oxygen. A polarized angle of 104.45° is present between the two hydrogen atoms as well as a distance of 95.84 pm between the nucleus of oxygen and any of the two hydrogen nucleus.

The water molecule or hydrogen oxide (H_2O) is found in all regions of the cosmos where temperature permits its formation and survival. It exhibits characteristics that are totally different from its oxygen or hydrogen components. Water has a high melting and boiling temperature as compared to hydrogen or oxygen. It has a hexagonal crystal structure in the frozen phase and a dipole moment-producing attraction between individual molecules in the liquid phase. Due to its polarity, water is a good solvent for a large number of substances, especially salts. Using surface tension and adhesion, it can raise against gravity by capillarity in a narrow tube. So water molecules have properties and qualities not found in their basic components, and these new attributes may be questioned for their permanence. Heat, chemical reactions, and electrolysis can destroy a water molecule and bring this compound to its natural entropic death. New H_2O molecules are born by igniting molecular hydrogen H_2 with an O_2 molecule or in a multitude of other reactions. But the lifetime of any individual in the complexity ladder is temporary and bound by its entropic fate just like Humpty Dumpty.

> *Humpty Dumpty sat on a wall.*
> *Humpty Dumpty had a great fall.*
> *All the king's horses and all the king's men*
> *couldn't put Humpty together again.*

7.3.3 Winning Cosmic Mix: Gravity, Black Energy, Electronuclear, and Biochemical

So it seems that our universe is filled to its ultimate limits with a winning cosmic mix made of gravitational, black, electronuclear, and biochemical energy bonds. We have no reason to believe that the laws of nature are different in some region of the cosmos. When we look at the Hubble Ultra Deep Field pictures from NASA, we find, except for the natural filtering from such a large distance, the same old and young galaxies expected from a uniform cosmos made up of a multitude of bubbles with different historical ages. The cosmological principle is saved as an average outlook over very large distances across many bubbles.

Most of the big bang problems [41], like the horizon, flatness, density fluctuation, singularity, and dark matter problems, melt away when the cosmos is viewed as a collection of bubbles of various ages that are individually born in a maxibang implosion/explosion whenever a critical mass equivalent to about seven thousand average-sized galaxies are gathered and gradually compressed and crushed to a tiny size at zero time and entropy under the force of gravity acting as a clock rewinder. The cycle is unceasingly repeated while every new version is produced by the homogenized admixture of mass energy from different neighboring bubbles, thus assuring cosmic uniformity.

BIBLIOGRAPHY

1 - Arp Halton, Catalogue of Discordant Redshift Associations, Montreal, Apeiron, 2003.

2 - Arp Halton, Seeing Red, Redshifts, Cosmology and Academic Science, Montreal, Apeiron, 1998.

3 - Beer Ferdinand P., Johnston Jr E. Russell, Eisenberg Elliot R., Clausen William E., Vector Mechanics for Engineers, Statics and Dynamics, Seventh Edition, New Delhi, Tata McGraw-Hill Publishing Co Ltd, 2005.

4 - Binney James & Tremaine Scott, Galactic Dynamics, Second Edition, Princeton, Princeton University Press, 2008.

5 - Carollo C. M., Ferguson H. C. & Wyse R. F. G., Editors, The Formation of Galactic Bulges, Cambridge, Cambridge University Press, 1999.

6 - Christensen Lars Lindberg, de Martin Davide, Cosmic Collisions, The Hubble Atlas of Merging Galaxies, New York, Springer Science 2009.

7 - Christensen Lars Lindberg, Fosbury Robert, Hurt Robert, Hidden Universe, Weinheim, Wiley-VCH Verlag GmbH & Co., 2009.

8 - Cohen-Tannoudji Gilles, Spiro Michel, La Matière-Espace-Temps, La Flèche, Éditions Gallimard, 1990.

9 - Diaferio Antonaldo, Editor, Outskirts of Galaxy Clusters: Intense Life in the Suburbs, IAU Colloquium 195, Torino, Italia, Cambridge University Press.

10 - Einstein Albert, Conceptions Scientifiques, Traduction de l'anglais, Manchecourt, Flammarion, 2004.

11 - Einstein Albert, Infeld Léopold, L'Évolution des Idées en Physique, Traduit de l'anglais, Manchecourt, Flammarion, 2004.

12 - Einstein Albert, La Relativité, Traduction de l'allemand, St-Amand, Petite Bibliothèque Payot, 1998.

13 - Greene Brian, La Magie du Cosmos, Traduction de l'anglais, Paris, Editions Robert Laffont, 2005.

14 - Hoskin Michael, Editor, The Cambridge Concise History of Astronomy, Cambridge, Cambridge University Press, 1999.

15 - Kembhavi Ajit K. & Narlika Jayant V., Quasars and Active Galactic Nuclei, An Introduction, Cambridge, Cambridge University Press, 1999.

16 - Lerner Eric J., The Big Bang Never Happened, New York, Vintage Books, 1992.

17 - Luminet Jean-Pierre, L'Univers chiffonné, St-Amand, Gallimard, 2007.

18 - Mészaros Péter, The High Energy Universe, Ultra-High Energy Events in Astrophysics and Cosmology, Cambridge, Cambridge University Press, 2010.

19 - Pascoli Gianni, La Gravitation, Paris, Presses Universitaires de France, 2ᵉ édition, 1995.

20 - Ratcliffe Hilton, The Static Universe, Exploding the Myth of Cosmic Expansion, Montreal, Apeiron, 2010.

21 - Reeves Hubert, Dernières Nouvelles du Cosmos, Paris, Éditions du Seuil, 1994.

22 - Reid I. Neill & Hawley Suzanne L., New Light on Dark Stars, Red Dwarfs, Low-Mass Stars, Brown Dwarfs, Chichester, Springer-Praxis, 2005.

23 - Schneider Peter, Extragalactic Astronomy and Cosmology, An Introduction, Berlin Heidelberg, Springer-Verlag, 2006.

24 - Séguin Marc, Villeneuve Benoît, Astronomie & Astrophysique, St-Laurent, Editions du Renouveau Pédagogique Inc., 1995.

25 - Serway Raymond A., Physics For Scientists & Engineers with Modern Physics, Third Edition, Philadelphia, Saunders College Publishing, 1990.

INTERNET ARTICLES

26 - Arp Halton, What is the mean redshift of the Virgo cluster?, 1988, Astronomy & Astrophysics, **202**, 70-76.

27 - Ashmore Lyndon, An Explanation of Redshift in a Static Universe, 2010, NPA, **Vol. 6** No 2, 1-6.

28 - Briggs Michael, Molecular Hydrogen Excitation in Gas Clouds, 2007, www.roe.ac.uk/ifa/postgrad/pedagogy/2007_briggs.pdf.

29 - Brownstein J. R., Moffat J. W., Galaxy Rotation Curves without Non-Baryonic Dark Matter, 2005, arXiv:astro-ph/0506370v4.

30 - Bruchon J., Fortin A., Bousmina M., Benmoussa K., Direct 2D simulation of small bubble clusters: from the expansion step to the equilibrium state, 2006, International Journal for Numerical Methods in Fluids, **00**, 1-29.

31 - Courteau Stéphane, Faber S. M., Dressler Alan, Willick Jeffrey A., Streaming Motions in the Local Universe: Evidence for Large-Scale, Low-Amplitude Density Fluctuations, 1993, The Astrophysical Journal, **412**, L51-L54.

32 - De Blok W. J. G., Bosma A., High-resolution rotation curves of low surface brightness galaxies, 2002, Astronomy&Astrophysics, **385**, 816-846, http://www.aanda.org.

33 - Folkman Judah, Hochberg Mark, Self-Regulation of Growth in Three Dimensions, 1973, The Journal of Experimental Medicine, **volume 138**, 745-753.

34 - Fraser Cain, Clouds of Hydrogen Swarm Around Andromeda, 2004, NRAO News Release.

35 - Galactic Rotation Curves, http://cosmology.berkeley.edu/Education/Essays/galrotcurve.html.

36 - Gregory Stephen A., Thomson Laird A., The Coma/A1367 Supercluster and its Environs, 1978, Astrophysical Journal, **222**, 784-799.

37 - Henry Richard C., Diffuse Background Radiation, 1999, The Astrophysical Journal, 516, L49-L52.

38 - Kassin Susan A., de Jong Roelof S., Weiner Benjamin J., Dark and Baryonic Matter in Bright Spiral Galaxies: II. Radial Distributions for 34 Galaxies, 2006, arXiv:astro-ph/0602027 v1.

39 - Marmet L., Gallo C. F., Comment on the use of Kepler's laws to describe galactic rotation, http://www.marmet.org/cosmology/nonkeplerian/galaxy.html.

40 - McGaugh Stacy S., Boomerang Data Suggest a Purely Baryonic Universe, 2000, The Astrophysical Journal, **541**, L33-L36.

41 - Meta Research, The Top 30 Problems with the Big Bang, 2002, Meta Research Bulletin, **11**, 6-13, http://metaresearch.org/cosmology/BB-top-30.asp.

42 - Ofer Yaron, The Galactic Environment of the Sun, http://www.tau.ac.il/~oferya/documents/The%20Galactic%20Environment%20of%20The%20Sun%20final.pdf.

43 - Pan Danny C., et al., Cosmic Void in Sloan Digital Sky Survey Data Release 7, 2011, arXiv:1103.4156v2 [astro-ph.CO].

44 - Percival Will J., et al., The 2dF Galaxy Redshift Survey: the power spectrum and the matter content of the Universe, 2001, Mon. Not. R. Astron. Soc., **327**, 1297-1306.

45 - Savin D. W. et al., The Impact of recent advances in laboratory astrophysics on our understanding of the cosmos, 2012, IOP Publishing, Reports on Progress in Physics, **75**, 1-36.

46 - Shiga David, Dwarf-flinging void is larger than thought, New Scientist, 2007.

47 - The Andromeda galaxy seen as never, at the 21-cm wavelength, 2009, http://www.obspm.fr/actual/nouvelle/oct09/m31.en.shtml.

48 - Types of Galaxies, http://www.universe-review.ca/F05-galaxy.htm.

49 - Van de Weygaert, et al., The Void Galaxy Survey, 2011, arXiv:1101.4187v1 [astro-ph.CO].

50 - Van Flanders Tom, Vigier J. P., The Speed of Gravity-Repeal of the Speed Limit, Meta Research, 2002, http://metaresearch.org/cosmology/gravity/speed_limit.asp.

51 - Weiss Achim, Elements of the past: Big Bang Nucleosynthesis and observation, 2006, Einstein Online, Vol. 2, 1019.

52 - Wikipedia, The free encyclopedia, Big Bang, http://en.wikipedia.org/wiki/Big_Bang.

53 - Wikipedia, The free encyclopedia, Black hole, http://en.wikipedia.org/wiki/Black_hole.

54 - Wikipedia, The free encyclopedia, Cosmic distance ladder, http://en.wikipedia.org/wiki/Extragalactic_distance_scale.

55 - Wikipedia, The free encyclopedia, Cosmic microwave background radiation, http://en.wikipedia.org/wiki/Cosmic_microwave_background_radiation.

56 - Wikipedia, The free encyclopedia, Dark Energy, http://en.wikipedia.org/wiki/Dark_energy.

57 - Wikipedia, The free encyclopedia, Dark matter, http://en.wikipedia.org/wiki/Dark_matter.

58 - Wikipedia, The free encyclopedia, Degenerate matter, http://en.wikipedia.org/wiki/Degenerate_matter.

59 - Wikipedia, The free encyclopedia, Doppler effect, http://en.wikipedia.org/wiki/Doppler_effect.

60 - Wikipedia, The free encyclopedia, Electromagnetic spectrum, http://en.wikipedia.org/wiki/Electromagnetic_spectrum.

61 - Wikipedia, The free encyclopedia, Extragalactic background light, http://en.wikipedia.org/wiki/Extragalactic_background_light.

62 - Wikipedia, The free encyclopedia, Friedmann equations, http://en.wikipedia.org/wiki/Critical_density.

63 - Wikipedia, The free encyclopedia, Interstellar medium, http://en.wikipedia.org/wiki/Interstellar_medium.

64 - Wikipedia, The free encyclopedia, Kepler conjecture, http://en.wikipedia.org/wiki/Kepler_conjecture.

65 - Wikipedia, The free encyclopedia, List of voids, http://en.wikipedia.org/wiki/List_of_voids.

66 - Wikipedia, The free encyclopedia, Local Void, http://en.wikipedia.org/wiki/Local_Void.

67 - Wikipedia, The free encyclopedia, Lyman-alpha forest, http://en.wikipedia.org/wiki/Lyman-alpha_forest.

68 - Wikipedia, The free encyclopedia, Main sequence, http://en.wikipedia.org/wiki/Main_sequence.

69 - Wikipedia, The free encyclopedia, Molecular cloud, http://en.wikipedia.org/wiki/Molecular_cloud.

70 - Wikipedia, The free encyclopedia, Neutron star, http://en.wikipedia.org/wiki/Neutron_star.

71 - Wikipedia, The free encyclopedia, Physics beyond the Standard Model, http://en.wikipedia.org/wiki/Physics_beyond_the_standard_model.

72 - Wikipedia, The free encyclopedia, Planck units, http://en.wikipedia.org/wiki/Planck_units.

73 - Wikipedia, The free encyclopedia, Quasar, http://en.wikipedia.org/wiki/Quasar.

74 - Wikipedia, The free encyclopedia, Redshift, http://en.wikipedia.org/wiki/Redshift.

75 - Wikipedia, The free encyclopedia, Soap bubble, http://en.wikipedia.org/wiki/Soap_bubble.

76 - Wikipedia, The free encyclopedia, Spiral galaxy, http://en.wikipedia.org/wiki/Spiral_galaxies.

77 - Wikipedia, The free encyclopedia, Stellar classification, http://en.wikipedia.org/wiki/Stellar_classification.

78 - Wikipedia, The free encyclopedia, Supernova, http://en.wikipedia.org/wiki/Supernova.

79 - Wikipedia, The free encyclopedia, Theory of everything, http://en.wikipedia.org/wiki/Theory_of_everything.

80 - Wikipedia, The free encyclopedia, Tully-Fisher relation, http://en.wikipedia.org/wiki/Tully-Fisher_relation.

81 - Wikipedia, The free encyclopedia, Vacuum, http://en.wikipedia.org/wiki/Vacuum.

82 - WMAP, What is the Universe Made Of?, map.gsfc.nasa.gov/ . . . / uni_matter.html.

83 - Wright Edward L., Age of the Universe, 2009, Tutorial/Age.

84 - Gallo C. F., Feng James Q., Galactic Rotation Described by a Thin-Disk Gravitational Model without Dark Matter, 2010, Journal of Cosmology, Vol. 6, 1373-1380.

INDEX

W

Z